Mitchell's Wings

© Johnny Carrington 2020

ISBN 978-1-5272-6303-1

The rights of Johnny Carrington to be identified as the author of this work have been asserted in accordance with sections 77 and 78 of the Copyright Designs and Patents Act 1988

Conditions of Sale

This book is sold subject to the condition that it shall not, by way of trade or otherwise, be lent, resold, hired out, or otherwise circulated without the publisher's prior consent in any form of binding or cover other than that in which it is published and without a similar condition including this condition being imposed on the subsequent publisher.
No part of this publication may be reproduced, stored in a retrieval system or transmitted in any form or by any means, electronic, mechanical, photocopying, recording or otherwise, without prior permission of the authors, singly or jointly.

A CIP catalogue record for this book is available from The British Library

Printed and Bound by BookPrintingUk
Coltsfoot Drive, Woodston, Peterborough

First Published in Great Britain in 2020
by Johnny Carrington Publications

Cover Design: Courtney Bryant @mashdesignstudio

Editor: Clive Hulme

All enquiries regarding rights associated with this play, including performing rights, should be addressed to:

Sophie Gorell Barnes,
MBA Literary Agents Limited,
62 Grafton Way,
London W1P 5LD

Tel 020 73872976
Email: Sophie@mbalit.com

johnnycarringtonpublishing@gmail.com

Photo Credits

Page (ii) Alex 'Chalky' Chalk playing Bob Doe at Middle Wallop capturing the audience with a monologue. N.B. Bob actually flew from Middle Wallop during the Battle of Britain. *cwphotos (Clive Weeks)*

Page (iii) Upper – a shot showing how the space was utilised at Middle Wallop and artistically 'framed' by the suspended aircraft. *cwphotos*

Page (iii) Lower- this was one of my favourite moments in the whole project; Chalky performing Bob Doe's eloquent thoughts about the beauty of the Spitfire... whilst actually touching one in the Solent Sky museum. *S Merredew*

Page (iv) Upper-the Spitfire appeals across ages and gender. Here Marie McDade performs Pat Viney's (née Maloney) words on seeing the Spitfire fly for the first time at the Solent Sky museum. *S Merredew*

Page (iv) Lower- The promenade type of performance allows a very intimate experience with the audience; aptly shown here at Middle Wallop in RJ's office. From the left. Marie McDade as Miss Cross (RJ's secretary), James Norton as RJ and Ian Wilson as Joe Smith (senior designer) *cwphotos*

Page (v) Rob Praine, as RJ, narrates to the audience at Middle Wallop. Again, it shows off well how the audience are surrounded by 'aviation'... the perfect set. *cwphotos*

Page (vi) A dress rehearsal at Oasis showing the use of multimedia *cwphotos*

Page (vii) RJ at work in his office. ***Solent Sky Museum***

Page (viii) Cast at Solent Sky performance *S Merredew*

Foreword by Johnny Carrington

If there could ever be something called a labour of love, then this project has been it! Having lived in Southampton all my life, I have always been proud of the Spitfire heritage connected to the city. This amazing plane first flew from the city's airport (Eastleigh) on March 5th 1936.

I remember visiting the Science museum as a youngster and being completely awestruck by this iconic machine hanging from the ceiling. As a drama teacher and playwright, I decided that I wanted to tell the story of the Spitfire not only to adults, but to a generation of children, some of whom had no idea about the city's connection with this iconic machine. Being a private pilot, ex-member of the RAF Reserves and general all-round aviation geek, it has even been suggested that I was 'born to write it'. Much to my family's amusement, if ever the sound of an aeroplane is heard overhead, I instantly start scanning the sky. But I also can't help but feel a huge sense of responsibility and privilege in telling this man's incredible story; I hope I've done him justice.

I consulted with Mark Wheeller, director of Oasis Youth Theatre and a colleague at Oasis Academy Lords Hill, along with Paul Ibbott, Head of Music, about my idea.

I subsequently put a proposal forward to Maskers Theatre Company, of which I am a member, to commission Mark, Paul and myself to write and direct a musical that would tell this story (Mark with the lyrics and Paul the music). They agreed that this could be an exciting opportunity to celebrate our local history whilst joining two respected theatre groups together - Oasis Youth Theatre and Maskers Theatre Company. One would provide the younger cast members and the other ... the not so young!

This really was a collaboration of talents and experience, and a massive thank you must go to the Maskers for funding the original project. I felt that for the published version, in order to reach as wide an audience as possible, I had to adapt the play and make it a straight drama rather than musical. However, I have left one song in and a link for the recording of this, along with the music can be found in the text. This can be used in any way you see fit.

Writing this play has been a real privilege. I have been humbled listening to the stories that have  been contributed and they all helped shape the piece, although regrettably there just wasn't the capacity to include them all. Stories of bravery, humility and modesty; from Peter Ayerst's 'amusing' recollection of being chased by twenty five Messerschmitts which he had accidentally

joined in formation over Germany, to Johnny Freeborn's riveting account of shooting down a Heinkel over London, to Bob Doe's moving testimony of how he couldn't bring himself to shoot down the helpless German fighter he had chased over the English Channel.

I had read biographies of and interviews with WW2 pilots, but talking to these gentlemen, first hand, knowing that they were barely twenty years of age during The Battle of Britain, has left me with nothing but admiration for them.

This project is not, however, just about pilots, or indeed R.J. Mitchell. It is about how the Spitfire itself has touched so many people and left such a lasting impression. Pat Viney and Joan Rolfe both recollect eloquently their experiences of the Spitfire; they talk with a sense of pride and fondness that is, I suspect, mirrored by many thousands of others... all of whom could tell their own story.

When this play was first performed, it travelled to three venues: my school (Oasis Academy Lordshill), The Museum of Army Flying at Middle Wallop and the Solent

Sky Museum in Southampton. Each venue had its own advantages and challenges, but each worked tremendously well using the 'promenade' style of performance. We felt that in order to perform at the latter two venues, with aeroplanes as a natural backdrop becoming part of a natural set, 'promenade' was necessary. The audience would be discreetly ushered around the different scenes by the cast and they became intimately involved in the story telling of the play.

Although to begin with I was slightly nervous of this style of presentation, without fail in each performance, at each venue, the audience twigged within the first five minutes how to 'behave' and it worked superbly. But this doesn't mean it can't be performed in a more traditional 'end on' setting.

As for the multi-media mentioned in the script, this can be adapted or even foregone completely. It could be that images/films/soundscapes are used just some of the time. Alternatively, it could be a very interactive performance

with film/images projected around the theatre (or on a suitably placed screen) in a way that submerses the audience in the 'Spitfire experience'. Could physical theatre be used to represent some of the images and 'moments' instead? A Crowd could easily be incorporated at different times to help portray particular scenes; this play really does afford the imagination of the director (and their technical crew) almost endless opportunities if they so wish.

I would like at this point to thank Dr. Gordon Mitchell (R.J's son) who sadly passed away in July of 2009. When I read his book 'Schooldays to Spitfires' as part of the research for this play I phoned him, and he immediately agreed to an interview and for me to use parts of his book in the play. His enthusiasm was infectious and I felt privileged at being able to chat with this gentleman about not only his father's creation, but also a story of determination and triumph against adversity that will hopefully educate and inspire. It was to my immense pride and gratification that the Mitchell family were able to travel to Middle Wallop and be guests of honour at the 'gala performance'.

It was also agreed with both Gordon Mitchell and Bob Doe that a donation would be made to Cancer Research and The Royal Air Force Benevolent Fund with every purchase of the script. On their behalf – thank you!

After the gala performance at Middle Wallop, I received a letter from Leonard Dickson, chairman of the Spitfire Society. The final paragraph has left a lasting impression on me.

> "It was a privilege to meet R.J. Mitchell's grandchildren and it must be a curious experience in seeing the portrayal of one's own relatives in quite emotive scenes. I must say that if I saw my relatives being shown in such a sensitive way, then I think I would have been very proud of your production. I do hope you can get the play to a wider audience in future – they would be guaranteed a very good evening indeed"

I hope that this play gives some insight into the extraordinary appeal of the Spitfire and how in the dark days of WWII, it was able to help inspire the nation... a legacy that lives on today.

Johnny Carrington – *Author / Director*

Wing Commander Bob Doe DSO, DFC & Bar

Bob Doe enlisted in the RAF in January 1939 and was posted to an operational squadron later in the year. He served with 234 and 238 Squadrons flying, initially, Spitfires and then Hurricanes. He was to become one of the leading aces of the Battle of Britain shooting down 14 aircraft, earning a Distinguished Flying Cross and Bar.

In January 1941, whilst flying a night sortie, the oil of his aircraft froze. As a result of his engine seizing he landed heavily at Warmwell on the snow-covered runway, breaking his harness and smashing his face against the reflector sight, almost severing his nose and breaking his arm. He was taken to hospital where he underwent

twenty-two operations. Returning to duty in July 1943, Bob flew against the Japanese in the Far East earning a Distinguished Service Order.

He retired from the RAF in April 1966 with the rank of Wing Commander and sadly passed away on 21st Feb 2010.

Upper - Bob Doe aged 19... a few months later he was fighting for his life in the skies above Hampshire.
Lower - A rather more tired looking Bob in front of his Spitfire Mk1 in 1940   ***Photos - Helen Doe***

Acknowledgements

Most importantly, I would like to thank all of the people I interviewed for this play. I'm sorry if not all the stories were included. It wasn't that they weren't worthy of inclusion, but rather that for dramatic reasons they didn't quite fit with the story I was telling. To be honest, I could have written three plays with the testimonies I had from: Dr Gordon Mitchell; Wing Commander Bob Doe DSO, DFC*; Wing Commander Peter Ayerst DFC; Wing Commander Johnny Freeborn DFC*; Sgt Joe Roddis, Pat Viney (née Maloney); Joan Rolfe (née Jagg); Sqn Ldr VR(T) Alan Jones MBE.

A big thank you to Bob Yeoman, who provided me with contacts for the pilot interviews.

The Maskers Theatre Company deserve huge thanks. It was their investment that made this project possible; I like to think that the reaction to the performances made it money well spent.

I had the privilege of not only working with some superb actors from the both Maskers and Oasis Youth Theatre (see cast list, page xiii), but the level of technical expertise delivered was amazing. Both by Danny Sturrock and the multi-media he created and also by Tony Lawther who co-ordinated all the "techies". Whenever I presented Tony with a challenge (and there were plenty of them), he would always come back with a very calm 'Yeh, we can do that'.

I would also like to say a massive thank you to Mark Wheeller and Paul Ibbott. Mark wrote the lyrics and Paul the music for the original production. It was a real privilege to be able to work on such a project with not only two talented colleagues but also close friends

A thank you must also be said to Alan Jones - and the staff - at the Solent Sky museum in Southampton. His 'can do' attitude prevailed in all our discussions, and this museum is a little gem. The same positive attitude can be said for the staff at the Museum of Army Flying at Middle Wallop.

Thank you to my editor Clive Hulme; your attention to detail and theatrical suggestions have helped immensely.

Finally, a huge thank you to my wife Sophie and my children Anna, Mia & Will. They had to put up with so many rehearsals and meetings; it couldn't have been done without their support... and watching Mia perform every night at the start of the show became a personal highlight.

Army Flying Museum Kentsboro, Stockbridge SO20 8FB

Solent Sky Museum Albert Rd S, Southampton SO14 3FR

Cast

Age suggestions are a guide, and many of the roles can be doubled

- Gordon Mitchell (son of RJ): adult of any age.
- RJ Mitchell as a boy (age 12-15)
- Herbert (RJ's father)
- Eric (RJ's brother) as a boy (age 12-15)
- Teacher: Adult any age
- Foreman: Adult male of any age
- RJ Mitchell as an adult (25-35)
- Flo Mitchell (30s)
- Mr Wainwright
- School receptionist
- Waitress
- Spectators 1 & 2
- Italian Pilot
- Race official
- Eric (RJ's brother, aged 30s)
- Ralph (worker at Supermarine)
- Henry Biard
- Miss Cross (RJ's receptionist)
- Ministry Official
- Narrator 1 & 2
- Joe Smith
- Barnes Wallis
- BBC Narrator (could be pre-recorded)
- Lady Houston
- Young Gordon (appears from ages 10 -15)
- Doctor
- Pat Viney (older)
- Pat Viney (age 14ish)
- Doris
- Aunt Connie
- Sir Robert McLean
- Bob Doe
- Young Bob Doe (12ish)
- RAF Recruitment Officer
- Derek (Pilot)
- Narrators 1, 2, 3 & 4
- Joan Jagg
- Peggy
- Mary
- Audrey
- Older Joan
- Foreman
- Jack
- John McKeown
- John's mum
- John's dad
- Doug
- Supermarine Worker
- Ken (Another friend of John)
- Dr Smitt
- Dr Smitt's secretary

Original Cast with Doubling

Rob Praine	Gordon Mitchell (R.J's son) - Narrator
James Norton	RJ Mitchell
Sarah-Jane Wareham	Flo Mitchell
Alex Chalk	Bob Doe
Sophie Agostinelli	Young Pat
Curtis Bain	Young Bob Doe
Paul Baker	Teacher, Foreman, Worker and Doctor
Stuart Barrow	John McKeown and Narrator
Kylie Boylett	Spectator 1, Younger Joan Rolfe and Narrator
James Bratby	Italian pilot, Jack and Narrator
Shaun Bridges	R.J Mitchell as a young boy and Ken
Carl Browning	Older Eric, Air Ministry Official and Mr McKeown
Mia Carrington	Doris
Steve Carrol	Sir Robert McLean
Sue Dashper	Lady Houston and Joan Rolfe
Bradley Holmes	Eric (RJ's brother), Narrator and Doug
Dylan Keenan	Young Gordon Mitchell:
Ruth Kibble	Mrs. McKeown and 'High Flight' Dancer
Marie McDade	Pat Viney & Miss Cross
Tasha Thomas	Mary and Narrator
Alan Watson	Barnes Wallis, Recruiting Officer and Narrator
Callum Watts	Schneider Official, Derek, Worker and Narrator
Ricolle Marie Williams	Peggy Moon and Narrator
Hayley Willsher	Spectator 2, Aunt Connie and 'High Flight' Dancer
Ian Wilson	Herbert Mitchell (R.J's father), Biard, Joe and Narrator
Becky Wiltshire	Audrey and Narrator

Additional members of the company as singers and dancers included: Sophie Agostinelli, Daisy Wheeller, Becky Wiltshire, Marie McDade, Carl Browning, Bradley Holmes, Shaun Bridges, Stuart Barrow, Ruth Kibble

Mitchell's Wings
The Story of the Spitfire

by
Johnny Carrington

Act I
Section 1: Early life

The stage is in darkness.

Pat Viney emerges into a spotlit area.

Pat Viney (older):	I had been evacuated to Swanage from Southampton because it was safe. Safe! By the end of the summer I had nearly been machine gunned by low flying Germans three times! I was evacuated with my friend Doris.
	Simultaneously, as Older Pat says "Doris" in a normal voice, Younger Pat offstage calls out "Doris!"
Young Pat (offstage):	Doris!
	She runs into an area which is suddenly spotlit as the spotlight on Older Pat fades to dim. Older Pat watches the following action.
	Come on catch up Doris!
Doris:	*(offstage)*: I'm trying.
	She enters the spotlit area, out of breath
YP:	Slow coach!
Doris:	You've got longer legs Patricia Maloney.
YP:	You're 'it'.
	Brief game of 'it' before both collapsing and lying down.
Doris:	Let's just lie here for a minute Pat, I'm boiling…

by Johnny Carrington

YP: It's so peaceful... picnic?

Doris: Cheese and cucumber or spam?

YP: Urgh! Spam?? Definitely cheese. *(Retrieving a toy monkey from her bag).* You'll have cheese too, won't you Monkey?

(As Doris sorts the sandwiches, Pat looks around then lies back staring at the sky, totally relaxed, Monkey on her chest)

I could go to sleep here Doris.

Doris: Me too.

Multimedia 1[1]

YP and Doris: Woah... Spitfire!

Both get up

YP: It's chasing that Messerschmitt.

Doris: Go on Spitfire!

YP: It's twisting and turning.

Doris: I can't see them... they've gone behind the cliff...

Pause

YP: There! They're climbing... the Spitfire's getting closer. You can do it!

Doris: Look, there's smoke coming out of the German.

YP: It's out of control... it's falling straight down.

Doris: It's crashed into the sea!

YP: I can't wait to tell Aunty.

---

[1] See Multimedia notes on page 85

They make to run off then freeze within their spotlit area. This fades to dim as Older Pat's spotlight comes up again.

Older Pat: It all seemed so detached from reality for us then... like a game. Someone had just died, but we didn't seem to realise that. I never could have imagined that having watched the first flight of a Spitfire four years earlier, I would later witness the Spitfire fighting the Battle of Britain.

Both spotlit areas fade to black.

Various cast members emerge into their own lit area(s) and narrate different lines of the following poem.

Oh I have slipped the surly bonds of earth
And danced the skies on laughter-silvered wings:
Sunward I've climbed, and joined the tumbling mirth
Of sun-split clouds - and done a hundred things
You have not dreamed of - wheeled and soared and swung
High in the sunlit silence. Hov'ring there
I've chased the shouting wind along, and flung
My eager craft through footless halls of air.
Up, up the long delirious, burning blue,
I've topped the windswept heights with easy grace
Where never lark, or even eagle flew -
And, while with silent lifting mind I've trod`
The high untrespassed sanctity of space,
Put out my hand and touched the face of God.

by Johnny Carrington

Gordon Mitchell: That poem was written by Pilot Officer John Gillespie Magee... a Spitfire pilot, aged just 19. I think my father would have found it rather amusing that a design of his could have elicited such emotion in a young man. Rather ironic then, that my father, RJ Mitchell, grew up at first not having a particular passion for aeroplanes at all. He was born in 1895 and the Wright brothers wouldn't leave the ground for another eight years and, like many small boys...

Male cast: ... and older boys...

Gordon: ... he was mad about trains. He was one of five children and grew up in comfortable surroundings near Hanley, in The Potteries. There was plenty of room in their grounds to play and keep themselves out of mischief.

Three boys, two girls playing, or just Reg and Eric for small cast

The boys, Reg and Eric, were actively encouraged to use their hands in a practical way and were given tools and materials by my grandfather to make things for themselves. But he was a stickler for perfection. Even a menial job, such as sweeping the floor had to be done thoroughly before it would pass inspection.

All cast except children inspect floor and make children do bits that were missed

Herbert: You missed a bit there... and there... and there... *(big pause – he seems to have finished)*... and there.

Gordon: My father was 14 when aviation first grabbed his attention. In 1908 Samuel Cody became the first man to fly in England. Reg and his brother Eric were so excited by this new and wonderful idea of flying that

	they started making their own model aeroplanes. They didn't have any instructions to follow so they simply made up their own designs using materials they found around the house and garden.
RJ(Boy):	Bamboo for the wings and fuselage…
Eric:	Glued on paper to cover it…
RJ:	Elastic stockings to hold it together…
Eric:	… and rubber bands to wind up the propeller.
	The boys then fly their aeroplanes around the stage.
	They are suddenly interrupted by mother
Female cast:	Boys have you seen my stockings anywhere?
RJ and Eric:	*(Hiding their aeroplanes)* No mother.
Gordon:	He was always inventing something at home to amuse his brothers and sisters. Most popular was his small sized billiards table. *(Eric exits and RJ mimes playing snooker)* Using stretched webbing fabric as cushions it provided hours of entertainment. But his father wasn't so happy…
Herbert:	Reg will never pass his examinations if he spends so much time playing billiards.
Gordon:	Taking all his exams in his stride he enjoyed school and devoured every scrap of information on aeroplanes he could find, and at a time when aviation was in its early stages, any new types and records being set grabbed the young lad's attention. Perseverance, a character trait displayed in abundance in his adult life was also apparent as a youngster. Approaching his teacher one day *(teacher steps forward)*…

by Johnny Carrington

RJ:	Sir, I want to read a novel
Teacher:	Excellent Mitchell. Who would you like?
RJ:	Sir Walter Scott.
Teacher:	Good choice. Which one?
RJ	All of them
Teacher:	I'm sorry?
RJ:	All of them sir... I want to read them all.
Teacher:	And he did. Ivanhoe, Rob Roy, The Lady of the Lake, Waverley... all of them, even the lesser known titles... quite a feat!

Young RJ puts on a pair of overalls

Gordon: At sixteen my father left School and worked as an apprentice at the locomotive engineering firm of Kerrs in nearby Stoke. One of his first jobs was to make the mid-morning tea for his fellow apprentices... and the foreman.

Foreman: *(Spitting it out).* It tastes like piss!

RJ: Sorry sir, I'll try harder tomorrow.

Gordon: The next day.

Foreman: Cup of tea Mitchell!

RJ: O.K. Sir won't be a minute.

Speaking to the other apprentices or audience

Don't drink the next brew... you'll see

He mimes going over to fill the kettle up, but this time he pees in it as well, makes the cup of tea and then brings it over

	Here you are sir... freshly brewed.
Foreman:	*(Taking a large mouthful the foreman savours it, swilling it round and round in his mouth. Eventually he emits a satisfied Ahhhh! turns to Reg and says)* Bloody good mug of tea Mitchell, why can't you make it like this every day?![2]
Gordon:	From top dog at school to grimy overalls and menial tasks... it was different to what he'd been used to.
RJ:	*(Moving to his home and interrupting his father reading the paper)* Father... It's not quite what I'd expected
Herbert:	What?
RJ:	The firm... I don't think it's my *(knowing glance at the audience)* cup of tea.
Herbert:	Really.
RJ:	I don't like it.
Herbert:	I beg your pardon?
RJ:	I said I don't like it!
Herbert:	Huh, you bloody well will like it!
Gordon:	So that was decided then! Dad was always rather impatient. His father had taught his boys to play chess and often enjoyed a game with them. *(Grandfather and RJ are playing chess)*
Herbert:	My style was rather different to Reg's... take your time and methodical. Reg, on the other hand, was very quick witted and couldn't understand why others couldn't keep up... Umm... there I think.

---

[2] This really happened!

by Johnny Carrington

RJ:	*(Responds almost instantly)* Your move father...
Gordon:	There is a long pause... a very long pause
RJ:	... Father?
Herbert:	I'm thinking *(Reg starts to pace).* I'm going... to... move... there... No, I'm not. *(Reg gets more agitated, maybe muttering under his breath).* There... check!
RJ:	*(Again responding instantly)* Over to you.
Herbert:	Now that's a puzzler. I'm really going to have to think about that.

(Reg's head drops and he groans)

Section 2: RJ Meets Flo

Gordon: My father impressed everyone at Kerrs with his talents. He wasn't a jack of all trades: he was a master of all, designing and engineering.

During this speech the boy RJ transforms into adult RJ. This can be achieved by the Young RJ handing his clipboard or another symbolic prop to the Older RJ.

He finished his apprenticeship and tried to join up and fight in the First World War. The government wouldn't allow it as they value his engineering skills. So, he started work in a local technical college... not what he wanted, but it did lead to a rather fortuitous encounter.

RJ is running to get to a lecture

RJ: Blast it, I am so late. *(He runs into Flo).* I am so sorry.

	(He scrabbles around, helping her pick up the files). I am a complete fool. I really should have been looking where I was going.
Flo:	*(Through slightly gritted teeth)* That's alright, no harm done.
RJ:	Please let me help you with them.
Flo:	It's alright, I only have to take them to that car over there.
RJ:	Do you want me to call your husband over?
Flo:	What? You cheeky so and so!
RJ:	Oh crikey, I'm so sorry… I'm not normally this stupid; I just thought you must be… I'll shut up now.
Flo:	*(With half a grin)* That's alright… again. People often make the wrong assumption. That's my car… and I like to drive fast… is that alright with you?
RJ:	Of course… sorry.
Flo:	Well thank you…?
RJ:	Reginald… err Reg… or call me RJ.
Flo:	Thank you. Cheerio Reg.
	Flo exits
RJ:	Cheerio
Mr Wainwright:	*(Calling from a distance).* Mr Mitchell? Mr Mitchell?… I think you must have dropped this file.
RJ:	Oh, that's not mine, that's…
	SFX: *car driving off*
RJ:	You've worked here a long time, Wainwright. Do you

by Johnny Carrington

	know who the lady driving that car is?
Mr W:	She works at the Junior School.
RJ:	Really?
Mr W:	Yes, she's the Head Mistress.
Gordon:	In no time, my father had tracked down the attractive lady… he was smitten.

SFX: *large motorcycle pulling up onto the gravel and sound of footsteps walking into the school*

RJ:	Excuse me, I'd like to see Miss Grayson if I may?
School Receptionist:	Certainly. Have you an appointment?
RJ:	Um, well no. I have a folder for her… she dropped it.
S.R.:	I can pass that on. *(She reaches out for the file)*
RJ:	Well I was hoping to see Miss Grayson

At that moment Flo walks into the office, at first she doesn't see RJ.

Flo:	Beryl, I've just been over to the technical college and I'm afraid all these files need sorting. A young oaf knocked straight into me.

(Beryl tries to indicate with her eyes that there is a visitor. Flo stops talking and turns slowly around)

Ah… Hello Reg.

RJ:	Hello Miss Grayson. *(Smiling)* I found another file… I thought it might be important, so I brought it around.
Flo:	Thank you. Very kind, you needn't have bothered to…
RJ:	Oh, no bother.

Flo: Well thank you. *(She goes to leave)*

RJ: Um, Miss Grayson, I was wondering if I could… um… have a word… if you don't mind?

Flo: Yes?

RJ: In private?

Flo: Alright, come through. *(RJ is taken through to the head mistress's office)*

RJ: Well, I hope you don't think I normally do this sort of thing… um… I was going to ask, if you weren't doing anything, which you probably are, so please don't worry, but I thought we might, if you wanted to that is, which you probably don't…

At this point Flo starts to giggle

Flo: I'm sorry Reg. What are you trying to say?

RJ: Just that if you wanted… would you let me take you out to dinner… or the flicks? *(There is a pause whilst Flo looks at RJ).* I'm sorry, this is stupid… sorry to trouble you. *(He goes to leave)*

Flo: I'd love to.

RJ: Really?

Flo: Really.

RJ: Wizard! *(He is beaming)* Shall I pick you up at around seven? There is a new Chaplin film out: 'The Rink'?

Flo: Lovely, I've been wanting to see that. I was wondering when it would get to Stoke.

RJ: Excellent… oh!

He stops suddenly and his face drops.

by Johnny Carrington

Flo: What?

RJ: It's just that, my car is in the garage at the moment and I only have a motor-bike.

Flo: Was that you on the Rudge?

RJ: *(Surprised that Flo knew it was a Rudge)* Yes it was... I mean is.

Flo: Well pick me up on that. Here's my address.

She writes on and hands him a piece of paper.

RJ is slightly wrong-footed by this.

RJ: Right... see you at seven.

Flo: Super, and don't worry about going too fast!

SFX: Motorcycle

Sequence - Motorcycle roars off with Flo clutching onto to Reg as they bank, are windswept and traverse a bumpy road.

Silent film jangly piano music. Flo and Reg sit side by side – as in the cinema – laughing along with the laughter of the other cinema-goers.

They return on the bike as above and it pulls up.

Flo: *(Dismounting and breathless)* Thank you Reg... I really enjoyed that... Chaplin is hilarious.

RJ: *(Also dismounting and 'parking the bike')* I know... can we do it again?

Flo: *(There is a slight pause)* I'd love to.

RJ: Same time tomorrow?

Flo: Same time tomorrow. Good night. *(She reaches across*

	and pecks him lightly on the cheek) Thanks again.
Gordon:	And so started their relationship. They enjoyed each other's company and the age gap - my mother was ten years older - was simply not an issue. In 1917 he saw a job advertisement for a personal assistant to Hubert Scott-Paine at the Supermarine Aviation Works, Woolston, Southampton. At last. Aeroplanes! He just needed to tell my mother…
	Transition music… the stage quickly has two small tables placed on it by the cast. RJ and Flo then sit at them.
Waitress:	Good afternoon, what can I get you?
RJ:	Tea for two and a couple of pieces of fruit cake.
Waitress:	Certainly.
Flo:	What's up Reg?… Come on tell me. You've been bottling something up all morning… Reg?
RJ:	*(Sighs)* It's this army thing… People think I'm a coward because I'm young and not in uniform. You saw how people looked at me when we came in.
Flo:	It's not your fault love, you tried twice to get in… and you are more important doing your engineering and if people have a problem with that, they need to get their facts right. *(She says this last line in a slightly louder voice).*
RJ:	But I'm not even doing that. I teach spotty fifteen-year olds for two days a week how to draw parallel lines… how is that helping the war effort? I need to be doing something Flo, this is driving me round the twist.
	He pauses

by Johnny Carrington

	There is a job that has come up... doing something for the war effort.
Flo:	Well that's wonderful.
	(Pause. Flo says enthusiastically) Isn't it?
RJ:	Yes... but...
Flo:	But what... Reg?
RJ:	It's in Southampton.
Flo:	Southampton?
RJ:	Working at Supermarine, building flying boats Flo. *(He starts to get more animated).* I mean this is more than I could have wanted, actually doing something directly to help the war... and with aeroplanes.
Flo:	Well you need to apply then. *(Hiding her slight disappointment).*
RJ:	Can I?
Flo:	Of course you can, you don't need my permission.
RJ:	We can still see each other can't we? I can ride up at the weekends.
Flo:	*(Flo pauses).* It's an awfully long way Reg.
RJ:	It won't take long, with my wages I can get a bigger bike... And there is nothing to say I'm going to get it. It's personal assistant to Hubert Scott-Paine... he's the general manager... seems like a real character.
Flo:	You'll get the job love.
RJ:	It's 1917, I don't know how much longer this war is going on for... and I just know that I won't be able to live with myself I don't do something before it finishes.

Flo:	Well I can't see it finishing any time soon. *(They sit in silence for a short while. Their tea arrives).*
RJ:	I will come up at the weekends though.
Flo:	Reg... *(Flo stops. He looks up from his tea)*
RJ:	What?
Flo:	If you meet someone else, please just tell me... be honest with me... please Reg? *(Flo is sounding a little upset, but still resolute).*
RJ:	I'm not going to meet someone else.
Flo:	Someone... someone... your own age. *(This has obviously been something that she has been thinking about).*
RJ:	I'm not looking for anyone else.
Flo:	But you probably will meet someone... younger...
RJ:	Why would I want to meet someone else... whatever age they are... it's never been an issue... Flo? Flo, look at me. ... It's nothing, it doesn't matter... I love *you*.
Flo:	*(Leaning across the table she takes hold of RJ's hands).* Go for the job Reg... I'll be here for you.
Gordon:	Not surprisingly he got the job and excelled at Supermarine. It was as if he had finally found his calling. He was challenged in new and exciting ways that tested his engineering and design skills: but despite this satisfaction that he now found at work, he was missing Flo.
	SFX: RJ's motorbike accelerating away
Flo:	It's lovely to see you Reg... a bit of a surprise, but lovely. How is work? How have you managed to get

by Johnny Carrington

	time off to come up and see me... let alone sort out a picnic!
RJ:	Hey, if I can design seaplanes now, I'm sure I can sort out a few sandwiches!
Flo:	They're even edible!
RJ:	*(Sarcastically)* Ha, ha, ha... They've given me a promotion at work.
Flo:	They **must** be short staffed!
RJ:	*(Laughing)* You can go off people you know. *(With a mock pompous voice)* I've even got my own office, more pay... *(More laughter - then changing tone).* It really feels like I am making a difference though: makes up for them not letting me join up.
Flo:	I've told you a hundred times, it doesn't matter that you couldn't join up. To be honest, I'm glad they wouldn't let you. *(There is a slightly awkward pause).* Boiled egg?
RJ:	Thanks. Actually, there was a reason for coming up. *(He is starting to become slightly awkward in his speech).* Well, like I said, I have got this promotion and they will be expecting more of me... and I'm not sure I will be able to come up here so often...
	(Flo suddenly interrupts him).
Flo:	Well thanks very much Reginald Mitchell! *(RJ starts to stutter as if trying to interrupt).* You take me out here on the pretence of a lovely picnic, just to tell me that you want to finish with me. *(RJ tries again to interject).* I know there is an age difference between us, but I honestly thought that didn't matter... not when two people loved each other.

	She starts to get up
RJ:	Where are you going?
Flo:	I'm going home. In fact, I'm going anywhere to be away from you!
RJ:	Wait, Flo.
Flo:	I can't believe I was getting all excited about this picnic and seeing you!
RJ:	*(Raising his voice as he stands up).* Just listen to me!
	Flo stops what she is doing in astonishment
	I'm not finishing with you.
	Flo is just looking at him
	I came here… I came here…
Flo:	You came here for what?!
RJ:	I came here… *(going down on one knee and producing a ring)* to ask you if you would do me the greatest honour and become my wife?
Flo:	*(Flo is obviously overcome).* Oh Reg!
RJ:	Is that a yes?
Flo:	I'm so sorry about having such a go at you, I feel such a fool.
Reg:	Well?
Flo:	I just assumed you were coming up to see me so you could let me down.
Reg:	Flo?
Flo:	What?

by Johnny Carrington

Reg: I'm waiting for an answer!

Flo: Yes! Yes! Yes! *(She rushes over and embraces him).*

RJ: Now I know the school means a lot to you, but I will be earning enough for both of us. We can rent a house near the river. Oh Flo... it will be perfect.

They kiss again

You would love Southampton Flo. I've got an office that looks over the water. The seagulls swoop and dive and it's the loveliest place.

Flo: But what about the family?

RJ: We'll tell them what we are doing now and get married, then we can move back down to Southampton.

Flo: Reg! *(They embrace)*

Gordon: My mother was a rock, a constant and also someone who shared his passions but she also understood his ambition and supported him throughout his career. Rapid promotion followed over the next couple of years... Chief Designer in 1919 and Chief Engineer a year later...

Section 3: The Schneider Trophy

Gordon: My father was starting to make his mark at Supermarine. An obsession began that inspired him, challenged him, frustrated him and drove him to the edge of despair... but ultimately brought him fame and money... his quest for the Schneider Trophy.

Multimedia 2 (Schneider Trophy) plays over the following.

The trophy aimed to encourage technical advances in civil aviation but became a contest for pure speed. The rules were simple. The race had to be flown over water, and each entrant had to be seaworthy. My father's interest began when Supermarine entered in 1919. Conditions, however, were far from ideal.

Spectator 1:	Look – there's the British aeroplane.
Spectator 2:	Where?
S1:	There on the right. (*Pointing*)
S2:	I can hear it, but I can't see anything - it's too foggy!
S1:	Just there.
S2:	No, that's the Italian.
S1:	Is it? Are you sure?
S2:	*(Sure)* Yes... *(Less sure)* I think...
S1:	I wish this fog would lift.
Gordon:	The result was declared void, however, when most of the aircraft couldn't navigate due to the severe fog. The Italians protested.
Italian Pilot:	(*Italian accent*) This is not fair, we made our way around the course.
Official:	I understand that, but you went the wrong way.
Italian Pilot:	That is but a minor detail, we finished it
Official:	But the wrong way sir.
Italian Pilot:	You are saying this because we are Italians

by Johnny Carrington

Official: I'm saying this because you only flew part of the course

Italian Pilot: Questa e una cospirazione, non abbiamo fatto niente di sbagliato. Voi non sarete affrontati dal fatto che siete stati picchiati dagl'Italiani !   *(This is a conspiracy, we did nothing wrong. You can't face up to the fact that you were beaten by the Italians!)*

Official: I beg your pardon sir?

Italian Pilot: Voi siete idioti e io mi lamentero! E non eancora finita!   *(You are idiots and I am going to complain. This is not finished!)*

Official: *(Patronisingly)* Thank you sir. Good afternoon

Gordon: Although the experience was rather frustrating, it inspired my father to design his own seaplane. However, the Italians were better prepared for the 1920 and 21 races... they won!
If they triumphed in 1922 the trophy would be theirs to keep for winning it three times in a row. Mitchell put all his efforts and energies into designing the Sea Lion II.

Multimedia 3 (Sea Lion)

Despite his drive and enthusiasm, he would worry constantly that something would happen to one of the pilots in an aircraft that he designed; often hardly bearing to watch the race. This affection for all his aircrew was deeply rooted and the friendships he shared with them would last his lifetime.

Seeing off entries from the U.S, France and the Italians, Supermarine won. It was to give him his first taste of real success.

The homecoming to Southampton was spectacular with thousands waiting to see glimpses of the winning pilot and trophy.

A crowd scene welcomes home RJ

My father was rather shy at the publicity and remained down to earth. One day my uncle visited.

RJ: Come on Eric, get your skates on.

Eric: Where are we off to?

RJ: The Red Lion.

Eric: The Red Lion?

RJ: Yes, come on.

Eric: Why the Red Lion?

Gordon: What's wrong with it?

Eric: Nothing, I just... well... thought you would rather... Doesn't matter.

RJ: When they arrived Eric could see why it had been chosen.

Ralph: Hello RJ

RJ: Hello Ralph, this is my brother Eric. Ralph works on the factory floor. Is Bill here?

Ralph: He's just getting the drinks. What are you having?

RJ: Two bitters please. *(Turning to Eric).* Bill is one of the fitters.

Eric: Do you know everyone in here Reg?

by Johnny Carrington

RJ: Just those from the works.

Gordon: Outside of work, people just felt comfortable with Dad, and he enjoyed their company where he was known to one and all as RJ. Despite his earlier victory, the American planes had developed and they won the next contest. In 1924 he had high hopes of regaining the trophy with his new monoplane the S4.

Multimedia 4 (S4)

Unique in its use of a single wing with no bracing wires, Dad hoped to regain the initiative. Speaking to one of his test pilots, Henry Biard, dad couldn't disguise the worry that accompanied him whenever he witnessed someone flying one of his planes.

Biard: Come on RJ, it can't be that bad, cheer up.

RJ: I'm fine Henry: you just concentrate on doing the flying.

Biard: I'm glad you said that RJ, because I was about to let it take off by itself.

RJ: Very funny... Look Henry, if anything goes wrong I have got my bathers on underneath. I can get to you very quickly.

Biard: *(Pretending to consider it seriously)* Thanks, but I'm not sure if that actually makes me feel more comfortable!

RJ: Just be careful.

Biard: Don't worry, my sense of self-preservation is still very strong. And besides, we don't want to worry the good people here with the thought of you suddenly stripping off.

Gordon: All seemed well, a series of high speed times were made and the initial hopes for the seaplane seemed to be realised. Then suddenly in the turn, the aircraft dropped, hit the water hard and broke up on impact. Biard, still strapped into his seat, was knocked unconscious and started to sink with his plane. The water around him turned dark as the plane sank deeper towards the sea bed. After launching the rescue boat, my father searched the crash area. Nothing could be found. Just a few bits of fabric and some oil floating on the surface. Suddenly the water bubbled, and Biard broke to the surface coughing and spluttering. He was dragged onto the launch and taken ashore.

RJ: Are you OK? What happened?

Biard: I don't know RJ. I think it might have been flutter in the turn again. She just fell apart and I lost control. Next thing I know everything is getting dark and the water has woken me up. I just undid my straps, held my breath and kicked for the surface.

Gordon: The cause of the crash was never positively identified. Most thought that the new wing design just wasn't strong enough. But despite the setback, my father was determined to race again for the Schneider trophy.

Lighting change. Back in the office

RJ: We can sort this. The basic design is sound. I'm sure of it... It's the fastest way for an aeroplane to travel. We just need to strengthen the wing.

Gordon: But this dedication and perseverance in sorting a problem meant that life wasn't always easy for my mother.

Flo: You're late back love.

by Johnny Carrington

RJ: I know… sorry.

Flo: You're working too hard Reg. You need some time off.

RJ: I can't Flo.

Flo: The hours you are putting in though.

RJ: *(Sharply)* I need to sort it. My design nearly killed Henry!… *(Going over to her).* My design will work, I just need more time.

Flo: I know Reg. *(Moving away).* I know you'll sort it out…

Gordon: Back at the office, he could be often seen with his elbows on the drawing board, face cupped in his hands trying to work out the next problem and woe betide anyone who came in at the wrong time.. His secretary Miss Cross found it difficult at first to gauge his moods.

Miss Cross: More letters for you to sign Mr Mitchell.

RJ: *(Taking all the letters and throwing them into the air).* There you are. *(Smiling charmingly)* I don't want anything more to do with letters. From now on you can answer almost all of them.

Miss Cross: Certainly Mr Mitchell.

Gordon: But slowly they developed a working relationship that ensured the best working environment for RJ and she would always vet possible interruptions.

Miss Cross: *(Speaking into the phone)* Good morning.

RJ: *(From off stage)* Come on Mitchell sort yourself out. Blast!

Miss Cross: You want to speak to Mr Mitchell? *(Door slams).* Better leave it until later.

Gordon: Despite his occasional outbursts, my father still got the best out of anyone working for him. He was a born leader, someone who would just as cheerfully say hello to a junior technician as to a V.I.P. He was also capable of standing up to the most senior of men from the Air Ministry if he thought they were out of line, not to mention if they breached his high moral codes.

Official: Now listen Mitchell, we need this to happen now. Not in six months.

RJ: But it's not ready.

Official: Well make it ready

RJ: I can't just magic a solution from thin air.

Official: Bloody Hell.

RJ: Don't use that sort of language in front of my Secretary!! *(The official exits rather chastened)*

Gordon: The Air Ministry agreed to formally back Britain's entry into the Schneider Trophy; it was to be a private venture no more... The 'High Speed Flight' was formed from experienced, serving, RAF pilots who would fly both Mitchell's aeroplane and their great rival's Gloster. My father had surrounded him with a team that would stay at Supermarine for many years.

RJ: *(To his team)* It is imperative we race in the 1927 contest. The Americans have won two in a row... a third would mean they keep the trophy.

Gordon: Flt Lt Webster was the test pilot. Dad was his usual bag of nerves: but this time it flew well. In his diary Webster wrote simply: Very nice. No snags.

BBC style Narrator 1: The 1927 contest took place in Venice. Surprisingly the Americans withdrew, saying it was not cost effective

by Johnny Carrington

	for the knowledge gained… as did the French, leaving just the British entries and the Italians.
BBC style Narrator 2:	After a delay for bad weather the expectant crowds witnessed the sea planes being towed from their hangers. Banners were everywhere expecting an Italian victory.
	The following race can be stylistically presented with physical theatre or simply narrated and/or augmented with a repetition of Multimedia 2 (Schneider Trophy). There is also a 'template' with clouds if the director wishes to create additional projections.
Nar 1:	Kinkead took off first in the Gloster and roared across the start line.
Nar 2:	Next was the Italian, Bernardi, and he went faster
Nar 1:	Webster was next in the S5 and started strongly.
Nar 2:	Then the second Italian, Guayetti, took off in the scarlet Macchi
Nar 1:	Worsley was the third Brit.
Whole cast:	*(Chorally)* It started raining.
	The cast start making rain sounds vocally or by tapping on the stage and continue throughout the following. If there are sufficient in the company they can split the rain effects and choral speaking among two groups.
Nar 2:	Bernardi dived down in a cloud of smoke towards the shore
Whole cast:	*(Chorally)* The others flew on.
Nar 1:	Suddenly Kinkead retired with engine trouble… everything was left in the balance.

Nar 2:	The other pilots could see what had happened but still they pressed on.
Nar 1:	Bernardi was in trouble: the engine was starting to run roughly... he put down on the water safely.
Nar 2:	All the British had to do was to keep flying.
Whole cast:	*(Chorally)* The laps were counted off...
Nar 1:	Although travelling over 270mph, they slowly edged through the seven 50km laps.
Nar 2:	Mitch, as he was now known to the RAF pilots, could barely watch.
Nar 1:	Eventually they passed the finish line... Flt Lt Webster won in the S5 with an average speed of 281.66mph...
Whole Cast:	... a new world record for sea and land planes!
Nar 2:	When he landed and found out the result, Webster said simply:
Male cast:	Now let's celebrate.
Gordon:	Back in Southampton a great welcome awaited the team and my father gave one of his rare speeches.
RJ:	I feel proud of being a member of the team which has brought back the Trophy. We gained considerable pleasure in the fact that we had won something for old England, because lately she has been losing sports laurels...
Gordon:	Sound familiar?
RJ:	... I think our victory will raise the Union Jack a little higher.

Actors respond with cheering.

by Johnny Carrington

There is a lighting change.

Flo: Goodnight Gordon

Young G: Is daddy coming home soon?

Flo: Yes… soon

YG: Can he read to me tonight?

Flo: He might be able to.

YG: It's the Jungle Book.

Flo: I know. We'll see what time he's back. Goodnight my sweet.

Gordon: Shortly after this the large company Vickers bought Supermarine, insisting that Mitchell stayed on as chief designer. However, a tragedy was to befall Mitchell that caused him much distress.

Joe: RJ, could you have a look at these drawings, I'm not sure we have got the scale quite right?

RJ: O.K. Joe. ..I see what you mean…

Miss Cross: Mr Mitchell, there is an important call for you.

RJ: I'll call them back.

Miss Cross: No, I think you need to take this.

RJ: Sorry Joe. Won't be a moment. *(Taking phone).* Mitchell here. *(Pause).* When? … Is there any chance?… Are you sure? What happened?… O.K. Thank you.

Joe: RJ?… RJ?… what is it?

RJ: It's Kinkead.

Joe: What about him?

RJ: He's crashed in the S5.

Joe:	What?... When?
RJ:	About half an hour ago. *(Mitchell hangs up)*
Miss Cross:	Can I get you anything Mr Mitchell?
RJ:	What? Um... no...
	(Pause)
Joe:	Is there any chance he could have survived?
RJ:	No... I don't think so. He was trying for the record. There was a break in the weather so he decided to go for it, dived down full speed... and just didn't pull up in time... Perhaps it was the S5? Perhaps I've missed something?
Joe:	There was nothing wrong with the plane RJ.
RJ:	You don't know that Joe. I want a full report, start it off.
Joe:	But why don't we wait until we...
RJ:	Now Joe! Right now.
Gordon:	In the days that passed my father worried that he was at fault somewhere.
RJ:	Perhaps I missed something. Get me the drawings again for the tail section.
Joe:	RJ, the aeroplane was fine.
RJ:	You don't know that for sure. He was killed flying something I created.
Joe:	But it wasn't your fault. The light was fading... you know it's difficult to judge the height above water at those speeds... it was being rushed through. They could have waited.
RJ:	What if I've made a mistake?

by Johnny Carrington

Joe:	You haven't RJ. I know it's difficult but you need to accept it... it was just a tragic accident.
Gordon:	It still remained unclear what had caused it, my father worked harder and harder: this increased the pressure on home life even more.

SFX: Phone dialling and then being answered.

Flo:	Southampton 458
Reg:	Sorry Flo, I'm going to be late again.
Flo:	Why?
RJ	I'm sorry... there is just so much... the accident has set us back. I think I'm nearly there though... Flo?.. Flo?
Flo:	Alright Reg... Actually, don't worry, I'm coming to pick you up.
RJ:	But what about Gordon?
Flo:	I will ask Mrs Woodman to pop around, I'm sure she won't mind. I'll be there in an hour.
RJ:	I might not be ready though.
Flo:	You said you wouldn't be long.
RJ:	I know, but I can't be specific.
Flo:	Well you can this time. One hour.
Gordon:	True to her word, an hour later she was there
	At the office.
Flo:	Hello there.
RJ:	Thanks, you really didn't need to you know.
Flo:	No trouble. Besides, we aren't going home yet.

RJ:	Maybe this isn't the right time Flo, I've still got rather a lot to do.
Flo:	No maybes about it… come on. I've made plans.
RJ	Plans?
Flo:	Yes. Gordon is being looked after, and we are going out.
RJ	Where?
Flo:	Surprise.
RJ:	It's really nice that you have been thinking about this, but I really do have so much work on.
Flo:	And I have been sitting around waiting for you for the last hour, so we are going. I'm driving.
RJ:	Not long though eh?
Flo:	Just get into the car Reg.
	Possibly blocks brought on stage (or maybe filmed and projected)
	SFX : Doors closing and engine starting. Gear changes and fast acceleration
RJ	Are we in a hurry then?
Flo:	Lost your nerve have you?
RJ:	No, but you are travelling a bit.
Flo:	You, of all people, worried about a bit of speed?
	SFX: Car can be heard going faster and faster
Flo:	*(Almost shouting)* I can't remember the last time I drove like this!
RJ:	Neither can I, if you don't slow down it might be the

by Johnny Carrington

	last time either of us do anything.
Flo:	Ha, the way you used to ride that motorbike, this must be like a donkey ride on Blackpool beach…
RJ:	*(Starting to get agitated).* Please, slow down a bit.
Flo:	Not yet
RJ:	Come on Flo.
Flo:	No chance.
RJ:	For Christ's sake, slow down!
Flo:	Why should I? You get your fun!
RJ:	Fun?!
Flo:	Yes, your fun.
RJ:	And what fun is that exactly? I work, if that's what you mean. But I wouldn't call it fun.
Flo:	Well at least you're working. I need to be doing something else.

SFX: Louder revs

RJ	Flo!

SFX: The car can be heard slowing to a more respectable speed

There is a pause.

RJ:	Why didn't you say?
Flo:	Why didn't you ask?
RJ:	I'm not a mind reader. And I've been so busy lately.
Flo:	You're always busy. And Gordon… he keeps asking where you are.

RJ	But the accident? I need to work out what went wrong
Flo:	Reg, I've been at the office when you were discussing it... the aeroplane was sound.
RJ:	You don't know that for sure. He was killed flying something I created.
Flo:	But it wasn't your fault. The report said the light was fading... you said it's difficult to judge the height above water at those speeds... Joe said it was being rushed through. They could have waited... no they should have waited.
RJ:	But what if it was me? Kinkead is a ... was a first class pilot.
Flo:	You haven't Reg. I know it's difficult but you need to accept it... it was just a tragic accident... No one could be more conscientious than you. Everyone knows how much you care for your pilots. If there was anything you could have done, you would have. And if there is anything wrong with the design you would have found it.
RJ:	Maybe.
Flo:	Not maybe... you would have. And... *(pause)*.
RJ	What... *(Pause)*... Flo?
Flo:	It's just that I need you, Gordon needs you... We need you.
RJ:	You've got me.
Flo:	Some of the time... but not much. I want a proper life together.
RJ:	But we don't have any money worries? I'm on a good

	wage and I...
Flo:	*(Getting agitated again).* Reg, you're not listening. It isn't the money... it's you I want. Just some of you.
	Pause.
	I miss our trips, the laughs we had, the fun.
	Pause.
	Sometimes I just want those times again. I don't care that we have more money now, I'd rather have more of you. I know that sounds selfish... and I feel guilty because it's the last thing you probably need to hear with what's happened... but I need something from you. I need more than I am getting. Please Reg?
	Pause.
RJ:	I'm sorry.
Flo:	It won't take much... just sometimes put us first. It gets lonely waiting for you.
RJ:	Alright Flo...
Flo:	I'm going to try and get another job. Not full time. Just something to help keep me active.
RJ:	You don't need to though. I earn more than enough for...
Flo:	I need something more than just waiting around. I need more of you and more of a life.
RJ:	I thought Gordon and me would be enough.
Flo:	And most of the time you are... but I need to keep busy. I'm meeting up with Janice from the club tomorrow. She knows someone who might have an opening in a local prep school. Just a couple of days a week.

RJ	I seem to be saying it rather a lot, but... sorry. I'm sure you'll get the job.
Flo:	Thanks... I'm not sure. But I need to try.
RJ:	Of course, and even if you don't... I will try.
Flo:	... Thank you.

SFX: Driving

RJ:	So, where to tonight?
Flo:	I think the new restaurant that has opened up just past the Bargate.
RJ:	Sounds expensive. *(She looks at him).* I'm joking!
Flo:	Just for that, I'm having a starter as well.

They both laugh a little... the awkwardness has gone

Gordon:	Vickers tried to appoint someone to help: someone who would later find fame and admiration as the designer of the famous bouncing bomb with the 'Dam busters'. Barnes Wallis was called in to work with my father... it didn't really work.
RJ:	What the bloody hell do they want to go and send me him for?
Joe:	He does have some quite new ideas on geometric construction
RJ:	I don't give two hoots. I don't want him working with me.

Enter Barnes Wallis... Mitchell immediately exits opposite.

BW:	Where did Mitchell go?
Joe:	I'm not sure, down to the factory maybe?

by Johnny Carrington

Barnes-Wallis exits, Mitchell enters.

RJ: Has he gone yet?

Joe: Only down to the factory

RJ: What a waste of my time!

BW: *(From off stage).* I can't see him, I'm coming back

RJ: I'm off! *(Exits)*

BW: *(Enters).* Did I just see him leave again?

Joe: Um... yes

Gordon: And so this went on. They were never in the same room at the same time. Eventually the chairman saw sense and moved Barnes Wallis on to something else. It had been a rather silly idea to try and make two geniuses work together. But at least now my father could concentrate on retaining the trophy.

Multimedia 5 (Schneider race 2 with commentary).

If not using the Multimedia file, the commentary is as follows:

"Flying the new Supermarine S5, Flt Lt Waghorn from the RAF High Speed flight successfully retained the prestigious Schneider Trophy for Great Britain. Flying an average speed of 328mph, Waghorn imposed himself from the beginning and never looked back. Well done sir!"

Gordon: Although he had feared the worst when, due to a rough running engine, he was forced to land. Disappointment soon changed to jubilation, however, for he had miscounted the number of laps he had completed... he

	had already finished the race. The attempts to win it for a third time in 1931 were hampered by the global depression and a reluctance for the British Govt to financially back it.
Miss Cross:	RJ, the French and Italians are competing, they've just announced it.
RJ:	Why won't our government? Don't they realise how close we are to winning it outright?
Miss Cross:	Can't we try and pull a few strings with the ministry?
RJ:	I've tried that already.
Joe:	*(Rushing in).* They've just announced it… they are pulling the plug.
RJ:	Are you sure?
Joe:	I've just had a call from the chairman. The government have withdrawn all funding.
RJ:	Idiots.
Gordon:	There was widespread condemnation in the press, but the government wouldn't budge.
	Pause
	Suddenly, however, an unexpected offer appeared.
Miss Cross:	Mr Mitchell, Lady Houston to see you.
RJ:	Lady Houston?
Miss Cross:	Shall I show her in?
RJ:	Um… yes of course.
Lady Houston:	Good afternoon Mr Mitchell.

by Johnny Carrington

RJ: Good afternoon Lady Houston, what can I do for you?

Lady H: Mr Mitchell, I can't abide the fact that you are so close to winning the trophy and don't have the money... will a hundred thousand pounds be enough?

RJ: I beg your pardon?

Lady H: A hundred thousand pounds... will it be enough?

RJ: Well... yes. Yes it would... but what, if you don't mind me asking, are you expecting in return?

Lady H: Nothing Mr Mitchell. We are a great nation and we need to win this trophy. Can you win it?

RJ: Yes, Lady Houston we can.

Lady Houston exits. RJ follows her out with his eyes. He stares at where she exited, looks around him as though needing confirmation of what has just happened then...

Joe! Joe! We can enter!!

Gordon: And so they raced.

Multimedia 6 *(Schneider race 3)*

The French withdrew, the Italians asked for more time following a crash of their new design, but the Royal Aero Club declined. And the race was flown uncontested in September 1931. Despite the worry that it was somewhat of an anti-climax thousands of people lined the shore in the Solent to see Flt Lt Boothman set a new world record and claim the Trophy outright for Britain.

Cast cheer

But perhaps the greatest legacy of the Trophy was the experience that my father had gained; experience that would influence his next major design, a design which influenced the path of British history...

Section 4: Illness

Gordon:	As a small boy growing up having a famous father that designed aeroplanes, there were some benefits. One of those was him taking me to the factory on a Saturday morning.
RJ:	Jump in boy.
Cast:	He did drive fast!
Young Gordon:	Bill said I could help him with metal work today dad.
RJ	O.K., just make sure you don't get in his way too much.
YG:	I won't, anyway he said it will be fine.
RJ:	Well today might be a bit different.
YG:	Dad, I won't be any trouble.
RJ	When we get there I want you to stay with Miss Cross for a few minutes.
YG:	Alright.
RJ	O.K here we are. Miss Cross, can I leave Gordon with you for a moment?
Miss Cross:	Of course Mr Mitchell.
	(RJ Exits)

by Johnny Carrington

	Here you are Gordon do you want a go on the type writer?
YG:	Thanks.
	Starts playing with it, making a mess
RJ:	*(Entering)* Right, all set?
YG:	What for Dad?
RJ:	How do you fancy a trip on the river in the company launch?
YG:	Really?
RJ	Absolutely... come on.
Boy:	Wizard!
Gordon:	It was the highlight of my week, and soon it would become a regular occurrence. Definitely a perk of being the boss's son... but I didn't realise it at the time.
	(Gordon looks on fondly as his younger self exits)
	At home we would often have the pilots around at our house, talking shop until the small hours. It certainly wasn't dull. In 1930 the Air Ministry put out a tender asking for a new fighter and my father, alongside his work on the Schneider trophy, began to design a new plane... he struggled... It proved to be underpowered and ungainly. Then, in 1933...
	RJ'S House
Flo:	Reg, I was thinking of arranging a match for us at the tennis club for Saturday week. Do you think you can play? Reg?... Reg? Are you alright? You don't seem yourself.

RJ:	Just a bit of an upset stomach that's all. *(Holding his abdomen).*
Flo:	Reg, what is it?
RJ:	I'm sure it will pass, stop fussing.
Gordon:	But it didn't, and a month later he was in some discomfort
Flo:	*(Seeing RJ holding his stomach obviously in some pain)* Love, you need to see a doctor.
RJ:	It will pass Flo.
Flo:	You've been saying that for weeks.
RJ:	I know.
Flo:	Sort it out then.
RJ:	Maybe.
Flo:	Reg!
Gordon:	No arguments there then… a couple of weeks later having seen his doctor he arrived home from work
	As he enters the house Young Gordon – now about 13 - is playing with a train set or similar period toy. He goes over and says hello. He then proceeds to the living room where Flo is listening to the radio.
	SFX: BBC News (background)
RJ	Hello love. Good day?
Flo:	*(Is distracted by the content and barely acknowledges him).* Hello… Reg, the BBC has said that Adolf Hitler is now chancellor of Germany. What do you think of him?

by Johnny Carrington

RJ: I think this chap could be trouble. I'm pretty sure that's why the Air Ministry put out tenders for the new fighter design. We're launching a major initiative to win the commission.

Flo: Reg, you don't think there will be another war do you? Not after what happened last time. There just can't be.

RJ: I'm sure it won't come to that. *(He winces in pain).* But we do need to be ready.

Flo: Oh Reg, I'm sorry, how did the appointment go? What did the doctor say?

RJ: Not much really. He sent me off to do a few tests and just to come back in a couple of weeks.

Flo: Good... I was thinking Reg... *(moves over to him, taking hold of his arms gently)...* we haven't been away for such a long time, and if you are going to get more involved in this new design, why don't we take a few days away? Gordon would love it and I'm sure the break would help you recharge the batteries a bit.

RJ is silent and thinking.

Please? Just a few days... the weather is turning warmer and we could go to the seaside? Torquay maybe? Or even just to Bournemouth? What do you think?

RJ: Alright, let's do it. You're right: I could do with a few days off. Gordon!? Gordon?!

Gordon comes in from the hall.

RJ: How do you fancy going on holiday to the seaside for a few days?

Young Gordon:	Oh wizard!
RJ:	*(Turning to Flo).* I'm going to really try to be here a bit more. *(Flo smiles and hugs him).*
Gordon:	So just before we were due to go on holiday to Devon, he went back to see his doctor:
Doctor:	Mr Mitchell, would you mind sitting down?
RJ:	What is the diagnosis then doc?
Doctor:	*(Pause)* ... It's not good, I'm afraid... I'm sorry, but there's no easy way of saying this... I think you have cancer.
RJ:	... Blimey... I see... Are you sure?
Doctor:	Not completely, I want you to see a specialist. But I think you need to be prepared for the worst.
Gordon:	It was confirmed. When he told my mother the news there was a difference of opinion as to what should happen next.
RJ:	The news isn't good Flo.
Flo:	Reg... what is it?
RJ:	*(Deep breath)* Cancer.
Flo:	Are they sure... but you're only 38... you shouldn't have it this young.
RJ:	It happens Flo.
Flo:	Oh Reg. *(Goes to him and holds him closely).*
RJ:	We can go on the holiday and forget about it for a few days.

by Johnny Carrington

Flo: What have the doctors said about it?

RJ: I've told them we need our holiday,

Flo: What did they say?

RJ: They want to operate.

Flo: When?

RJ: They said... look I don't want you and Gordon to be put out.

Flo: Reg.

RJ: They want to do it straight away but...

Flo: There will be no 'buts' Reg... we are not going on bloody holiday! You are getting that operation done straight away, and don't you dare argue!

Gordon: The operation was successful, but distressing. He had his rectum removed and a colostomy bag fitted. My father demanded to know what the prognosis was.

Doctor: Mr Mitchell, the operation was successful. With some adaptation, patients can live quite normal lives.

RJ: What about the future though?

Doctor: We removed the cancer, which was the main problem, but it is difficult to really diagnose precisely what will happen.

RJ: I don't need flannelling Doctor, I need to know what my chances are... they're not good are they?

Doctor: It's too early to say.

RJ: Are they!?

Doctor: ... I am afraid to say that there is a risk that the cancer will reoccur. If you survive for four years without it

	returning then life will start to look a lot better.
RJ:	... Thank you.
Gordon:	The operation had been in the summer of 1933. He could easily have retired then and said 'I will enjoy what little time I have left, I am not going back anymore'. Supermarine would have made sure he was financially secure - that could easily have happened, but the result would have been -no Spitfire. He didn't and a few months later he was back at work. His first attempt at the new fighter had been largely unsuccessful, but he had new ideas.
RJ:	Now listen Joe, we need to get our heads around this new spec. I want to think differently. We don't want a fixed undercarriage, we want retractable. And the engine... it's not right.
Joe:	I know RJ, it's just not reliable enough.
RJ:	Rolls Royce has a new power plant... the Merlin. Why can't we fit that instead?
Joe:	I think it can be done.
RJ:	Let's try it. And the wing, I want to change the shape.
Joe:	To what?
RJ:	I want it to be more elliptical, Shenstone has done some superb work here... and make it thinner but thick enough near the root to house the undercarriage.
Joe:	That will be more difficult to mass produce RJ.
RJ:	I know, but the manoeuvrability will be much better. And it is supposed to be a fighter... it will travel faster. At the moment our bombers are faster than our fighters... This could be something really special Joe, I

by Johnny Carrington

	want every effort put on this.
Gordon:	Perhaps spurred on by the knowledge he was very ill, my father worked harder at the new design. To fully inform him on all developments and to understand everything, he even obtained his pilot's license… All the while no-one at work knew he had cancer, just mum and me. He was determined to do everything to cope with it. I remember seeing him at work in the garage one afternoon.
Young Gordon:	Dad, what are you doing?
RJ:	Just working on something Gordon.
YG:	What dad?
RJ:	Something to make things a bit easier.
YG:	What do you mean?
RJ:	My… colostomy bag Gordon…
YG:	Oh.
RJ:	I think I have worked out a better design.
YG:	You are still coming to visit aren't you dad?
RJ:	Visit?
YG:	School dad. You haven't forgotten have you?
RJ:	Of course not.
YG:	Are you sure you will be alright?
RJ:	Absolutely… wouldn't miss it for the world

SFX: Car

RJ and Flo are in a car travelling.

Flo:	It's lovely that you want to visit Gordon at school, but I am really not sure about this cricket match. You need to be taking it easy; you know what your doctor said.
RJ:	It's fine Flo, I promised Gordon I would be there, and I'm not going to let him down.
Flo:	But he'll be just happy to see you, you don't need to do the cricket.
RJ:	I have been feeling pretty well recently though… I won't push myself too much, I promise.
	SFX: Car pulls up on a gravel drive. The sound of doors being opened and closed
Young Gordon:	*(calling from offstage)* Dad! Mum!
Flo and RJ:	Hello Dear/How are you?
YG:	*(entering)* Have you brought your whites dad?
RJ:	Of course I have.
YG:	Great, the teachers are being really cocky about beating the parents. I'll show you around and then it's lunch. Cricket is this afternoon at 2.30.
RJ:	Lead on! *(Aside)* I'll be fine Flo.
	SFX: Cricket match
	A choreographed sequence – possibly in slow motion - where RJ hits three sixes and is then caught out.
	The stance, the run up, the strike, the crowd, the reaction to each six and the final catch.
Cast:	Six! Six! Six!

	Ahh...!
YG:	Thanks so much for coming up... and well done for the cricket dad... they'll be talking about you tonking Jeffreys around the field for months.
Flo:	Do you want to drive Reg?
RJ:	You can drive love. *(Flo is slightly perplexed, but walks around, stopping next to Gordon to give him a hug before she gets in).*
Flo:	Take care love and write if you need anything.
YG:	Thanks mum.
RJ:	Cheerio Gordon.
YG:	Bye dad.

SFX: Car accelerating away

RJ grimaces, finally betraying the pain he is feeling.

Flo: Oh Reg, I knew you were pushing it... do you want me to stop? *(RJ shakes his head).* Oh love.

She puts out a hand and strokes his leg.

Now Reg, you are not going in tomorrow. Joe can look after the office for a day or two.

RJ just nods.

Gordon: By the beginning of 1936 the prototype Spitfire, just called K5054, was nearly ready for its first flight. Pushing himself relentlessly, there was a rather telling entry in his diary for 31st December 1935. "Fairly severe back ache. Rather concerned re future".

Act II
Section 5: First Flight

Multimedia 7 (blueprint)

Gordon: In February there were the first engine runs, and then on March 5th the prototype was prepared for its first flight. The whole factory felt a sense of pride that something special was being created… and they all felt part of that team. But it wasn't just the workers who were there.

As older Pat narrates, the action takes place elsewhere on the stage.

Pat Viney (older): I was lucky enough to be there to see that first flight… and what an impression it left with me. My father worked at Supermarine and his job was working on the aeroplane's engine, so my aunt knew its first flight was about to happen. All I knew at the time was I was being taken on another day trip with her. I asked her:

Young Pat: Where are we going Aunt Connie?

Aunt Connie: You'll see Pat.

YP: Come on tell me.

AC: It's a long journey.

YP: Where though?

AC: We're going to see the new plane Daddy's been working on take its first flight at Eastleigh.

by Johnny Carrington

YP:	Wonderful!
Pat:	It was the longest journey I had been on.
YP:	I feel sick Auntie Connie.
AC:	Look out of the window dear.
	(She looks out of the window. Pause)
YP:	I still feel sick… are we nearly there yet?
AC:	Soon, have a barley sugar that will make you feel better.
Pat:	Luckily my aunt had a good supply. Eventually we reached the airfield: I'd never seen one before.
YP:	Wow! It's massive Aunty Connie. There is so much grass: it goes on for miles and miles.
AC:	Come on let's get a spot where we can see.

Multimedia 8 (Maiden flight)

Pat:	Gradually people arrived and there was a lot of chitter chatting going on – anticipation of what would happen. It was more or less full when suddenly the hangar doors opened. The chatter stopped and instead you could hear 'She's coming out'… 'Look it's coming out'… and it just was… it came out gradually, slowly under its own power, like driving a car. People started to clap, then cheer and wave.
YP:	Look there's daddy.
AC:	So it is… wave.
YP:	Daddy!... he waved back.
Pat:	It moved to the centre of the airfield, the crowd full of anticipation. She lined up, and then we heard the roar of the engine as she gathered speed, moving faster

and faster over the grass, then left the ground and soared gracefully into the sky.

SFX : Merlin engine take off sound

It was the most amazing thing I had ever seen... I could never have imagined that four years later I'd be watching Spitfires again... this time fighting with German planes over Swanage. A couple of days later little gifts were given out to commemorate the first flight. My father got me a small toy monkey... I still have it and it reminds me of that amazing day... all those years ago.

Gordon: Over the other side of the airfield my father was, as ever, relieved that the first flight was safely over. He rushed over to Mutt Summers the pilot.

RJ: How was it Mutt?

All: I don't want anything touched.

Gordon: Much has been made of that quote... did he mean it was perfect and didn't need anything being done to it? Or did he mean, he needed everything left as it was before the next flight? It probably meant there were no major snags which needed fixing. But that quote has added to the mystique of the Spitfire. Barely three months after that first flight there was some very good news: an order for over 300 of the new design.

RJ heads back to the office. Joe catches up with him.

Joe: Now RJ, you haven't forgotten that Sir Robert is here to talk about the name for K5045.

RJ: *(resignedly)* I know Joe.

by Johnny Carrington

RJ enters the office where Sir Robert is waiting

Sir Robert: Morning Mitchell, morning Smith.

RJ: Morning Sir Robert.

SR: I have come up with a name you will rather like.

RJ: Really sir?

SR: Spitfire.

RJ: Spitfire?

SR: Spitfire... my youngest Ann, just running around everywhere gave me the idea... spitting fire everywhere she went... Spitfire... Let me know what you think? Excellent, well must dash. Maybe have lunch on Tuesday. Cheerio Smith.

Sir Robert exits

Joe: Cheerio Sir... What do you think RJ?

RJ: Huh, it's a bloody silly name!

Gordon: And so the Spitfire legend began to grow... When dad was first working at Supermarine in the 1920s, people were being born who'd become future pilots and ground crew, office workers and engineers... people who shared something. People who would have their lives touched by the Spitfire... people like Bob Doe.

Bob Doe: One of my first memories is of going to Leatherhead Central School. It was about five miles from home and about five hundred feet lower. Going was wonderful: downhill all the way on my bike.

Young Bob:	WOOOAAHH!!!! *(crossing stage on bike from wings into opposite wings, feet spread wide – no need to pedal!)*
Bob Doe:	Coming back was hard work.
	(Re-enters pedalling uphill furiously, or pushing his bike).
	I would often able to hang on to the odd lorry and get a tow. Some of the lorry drivers would wave to me as I let go. Then one day on the way home…
	Multimedia 9 (biplane coming in low accompanied by sound)
YB:	WOW!! That is low!
Bob Doe:	An RAF biplane was force landing in the field next to me.
YB:	'Scuse me mister.
	In the original production we had Bob prompting an audience member to answer Young Bob's questions as though they are the pilot. The audience member was picked out at random by the actor playing Bob and there was never a refusal.
	Alternatively, have another actor playing the part of the pilot of the aeroplane which has force landed.
	In the question above Young Bob is trying to elicit the answer Yes lad? *or similar.*
YB:	Can I have a look at your aeroplane? (Pilot: Of course, help yourself).
	Young Bob enacts the following
Bob Doe:	I was able to walk around it, touch it and feel it. To

me it was the start of the mystery of Aviation. I was late home that day and thoroughly ticked off. But the lad who had seen that old biplane force land next to him on the way home from school was now fascinated by aviation. Growing up, I had made models like any other boy and would re-enact air battles from the First World War.

We see young Bob playing with a model and then transform into a young man – the model is passed from one actor to another.

As a young man, I still had that curiosity for aeroplanes and...

Section 6 : Entering RAF Service

Bob Doe: ... barely a year after the first flight of the Spitfire, I decided that I would learn to fly, I joined the Reserves and they taught me. However, I wanted to be an officer in the RAF – but had left school at 14 – the Air Force want you to have quite a good education and I didn't ... I went to London for my interview.

RAF Recruiter: Mmm... not very clever at school were you Doe?

Bob: No sir.

Recruiter: Mmm. *(Long pause).* Left school at 14?

Bob: Yes sir.

Recruiter: Mmm *(Long pause).* Any sports Doe?

Bob:	I play cricket. And I ran in the National Juniors for hurdling.
Recruiter:	Ahh. That's better. You will have to sit an exam because... well... you didn't pass many at school.
Bob:	O.K. sir.
Recruiter:	Here, read and learn this... just the paragraph at the back... the formula for the movement of the arm.
Bob:	Just that paragraph?
Recruiter:	You can learn it all if you like, but your academic record suggests you won't cope terribly well... so *(hinting heavily)* just the paragraph at the back.

Starts exiting slowly

Bob:	Right... thank you sir. And so I read it, it came up in the exam ... and I passed!

RAF Recruiter turns just before leaving the stage and does a heavy wink towards Bob

And that's how I joined the RAF. Two years later, and now a regular squadron officer, my life was changed by a new arrival at our station...

He blends into the squadron scene

Derek:	Bob, quick, there's something coming in.
Bob:	Derek, I am perfectly comfortable sitting here. Why would I want to go to all the effort of standing up, just to watch another clapped out Blenheim flying a ropey approach?
Derek:	It's not a Blenheim.
Bob:	I'm still not moving.

by Johnny Carrington

Derek:	It's a Spitfire.
Bob:	What?
Derek:	A Spitfire.
Bob:	If you are fibbing Derek!
Derek:	Honestly, look.
Bob:	*(Getting up)* Crikey. What's she doing here? We looked at it. We sat in it. We stroked it. I fell in love with it. It was the most beautiful thing I had ever seen... it really was. And then the following day fifteen more turned up.
	I got into this huge aeroplane, huge to me anyway, a big nose up front and yet the first time I took off, we had no pilot's notes or anything like that, I felt at home. It was the most comforting aeroplane I have ever sat in. You didn't have to turn it, you just had to think about it and it turned for you. It had no vices. It did everything you wanted it to do. It was just a dream, I loved it... and now... we were a Spitfire squadron.

Section 7: The Battle for Britain

Gordon:	Although created in peace time, it was in war that my father's creation was to have its finest hour
Nar 1:	The battle for France was swift but disastrous. Within a month of invading, Germany had pushed back the Allies.

Nar 2:	Only a miraculous escape at Dunkirk saved what was left of the British Army.
Nar 3:	Now Hitler turned to face Britain. His first task was to knock out the RAF. With air superiority, Germany could invade.
Nar 4:	It was left to a few hundred Allied airmen, heavily outnumbered, to repel the German air force... the Luftwaffe. The two aircraft tasked with defending Britain was the Hurricane and the Spitfire.
	The following can be soundscaped by the cast or SFX specifically created. The first part of this should be prefaced by radio calls made by pilots to ground controllers and to each other. This is followed by the sounds of battle as well as the sounds of a pilot being shot down. It needs to emphasise the mayhem and danger.
Bob Doe:	I was the youngest pilot on the squadron... and the most scared. Our first scramble was against 200 aircraft coming over the Swanage area. We took off from Middle Wallop and did everything wrong. We knew nothing about fighter tactics; we went up in tight formation which was something you should never do in a battle area. We flew at the same height as the bombers, which was bloody stupid... you want to get above them. We proceeded to patrol up and down the sun – how stupid can you get. Before we knew what had hit us the last section had lost two of its pilots and its leader... dead. Eventually we found the enemy – the sky was full of black crosses – how we got into the middle of them I will never know – I found myself sitting behind a German 109. We started turning and it started shooting at me so I shot back and shot it down. In that second ... I became a

by Johnny Carrington

	fighter pilot. I realized the power of this beautiful aircraft and I don't think I was scared again. When we came back I found that we had lost three out of the twelve. That night I went to bed early and thought about how I was going to stay alive. You needed to see the enemy first... so keep looking. If you flew in a straight line for more than ten seconds you would probably be dead – so keep turning. These were the lessons I taught myself... because there was no one to tell us what to do.
Gordon:	Back in Southampton, the Supermarine works in Woolston employed hundreds of people to keep pace for the demand of Spitfires needed by the RAF. It wasn't just men that were needed. Joan Jagg worked in the accounts dept...
	At the Supermarine factory, Southampton
Joan:	Ok then girls, whose house will it be?
Peggy:	You can come round mine if you like. Dad's out and mum will make us cakes.
Mary:	How about you Audrey? Will your fiancé let you out again?
Audrey:	Yes of course.
Peggy and Joan:	Two weeks in a row?
Audrey:	Geoffrey isn't like that. He knows that we have our Tuesday night meetings and he's not at all jealous.
Joan:	Seven thirty?
Peggy, Mary and Audrey:	See you then.

Older Joan:	Peggy was my best friend and we worked together at Supermarine, although we were on different sites. All the girls were friends at school, and after leaving we would meet up every Tuesday for a girls' night in. We would call it 'The Hen Night', although we rarely drank. Even when the war began, we would make it around to someone's house.
Audrey:	I've made some cakes.
Peggy:	Lovely, where did you get the flour from? I've used up my ration already.
Audrey:	I couldn't get any, so I've used grated potato instead.
Mary:	And the sugar?
Audrey:	Grated carrot and just a touch of paraffin.
Joan:	What?!
Audrey:	We have to make do.
Mary:	I'd rather not 'do' to be honest
	Air raid siren goes off
Peggy:	Perfect timing. Come on let's move into the Anderson shelter.
Mary:	I hate those places, they smell horrible.
Joan:	Better the smell than caught in a house that's just been bombed.
Audrey:	Geoffrey said that a friend of his was killed in an Anderson shelter and the house was left completely untouched.
Peggy, Joan and Mary:	Audrey!!!

by Johnny Carrington

	They move into it by simple change in lighting and position. We also hear bombs and ack ack
Joan:	This is depressing
Mary:	And smelly.
Audrey and Mary:	*(singing)* There'll always be an England...
Peggy and Joan:	NO! ...
	Everyone is quiet for a bit.
Audrey:	Anyone for a cake?...
	Silence
Peggy:	I think Dad fixed the wireless.
	Starts to tune it in and picks up Winston Churchill's speech 'We will fight them on the beaches' (available in the downloadable files). They listen and it fades into the background, then away completely, as:
Older Joan:	And that's where I was when I heard Churchill's famous speech: Stuck in Peggy Moon's Anderson shelter... The following Saturday was Audrey's wedding. It was lovely... Peggy and I walked home together. The searchlights were moving across the night sky, but there were no air raids, it was somehow eerily beautiful.
	*Lighting to show search lights or **Multimedia 10 (Searchlights)***
Peggy:	Do you think we will get married Joan?
Joan:	Course we will Peg... We're just waiting for the right man.

Peggy:	That could take a long time then.
Joan:	I'm sure it won't. Audrey's got Geoffrey and Mary has started seeing Bill.
Peggy:	I know, but with this war on more of the men are moving away. I know that sounds so selfish doesn't it?... I don't want to be left on my own.
Joan:	You won't be. I saw the way Geoffrey's brother was looking at you during the reception. And he asked you if you wanted more cake.
Peggy:	*(Laughing).* What is that supposed to mean?
Joan:	You're as good as married!
Peggy:	You're mad Joan Jagg... seriously though... I do want to meet someone.
Joan:	You will Peg, you will.
Peggy:	I hope so. Nothing ever happens to me.
Joan:	Here's my road. See you at Mary's on Tuesday.
Peggy:	See you Tuesday.
Older Joan:	Tuesday came and at work we could hear the bombs being dropped again. Stupidly I went over to the window to see them.
Foreman:	Come away from those windows... now!
Joan:	Are we going outside again?
Foreman:	No, just to the basement.
	Soundscape of bombing raid underscoring the following.
Older Joan:	We could hear the attack happening... a couple of times we felt the walls shake... and then it stopped.

by Johnny Carrington

	The all clear was sounded.
Foreman:	Come on Joan... let's go
Joan:	There's so much dust and glass. And that smell... it's horrible.
Foreman:	Just have a look and see if you can still get your things. I'll see you outside... I think we missed the worst of it.
	Foreman exits
	A young lad enters
Joan:	Jack, have you come to give us a hand?
Jack:	No, I've been on the Woolston site as a stretcher bearer.
Joan:	Oh Jack. I'm sorry... are you alright?
Jack:	I think so... but it's not good though. They got a real pasting down there.
Joan:	How many?
Jack:	I don't know... quite a few...
Joan:	You poor thing.
Jack:	... It's Peggy...
Joan:	What about her... Jack? What about Peg?
Peggy:	*(Walking calmly to the back of them or to one side of the stage)* I don't want to be left alone.
Jack:	She was... was... caught in one of the shelters. It took a direct hit... I'm sorry Joan... I knew she was your mate, so... I wanted to tell you... I'm really sorry.
Peggy:	Nothing ever happens to me

Joan: Are you sure?

Jack nods and Joan sits down her head in her hands

Older Joan: Peggy was only 20. My best friend… it had all seemed… not quite real… until then… she was the only woman killed in the bombings of Supermarine… I still think about her and those words she said…

Peggy: Nothing ever happens to me.

Older Joan: Recently there was a memorial at Woolston and a plaque unveiled. It was lovely to see some of my old friends from the Tuesday Hen Nights… we laughed again about some of those evenings… I was proud of my time at Supermarine, to think we all helped make the Spitfire was a wonderful thing… but it also brought back some sad memories… and I left a pink rose… for Peggy.

Gordon: Despite the savagery of air combat and the killing that occurred on a daily basis, Bob displayed a mercifulness that few would have thought possible in the heat of battle.

SFX: MM13

Bob Doe: We were in a big dogfight over the Isle of Wight. An aeroplane came across my sights and I had a snapshot at it. It turned over and dived down through the mêlée towards the sea so I pulled up and over. I wanted to keep an eye on him, and when he got down to the sea he just turned and headed for home… I thought – you're not going to get away with that so I went after him – it took me a hell of a long time to catch him up but eventually I did and I took a shot at him. His hood came off, his wheels came down and his engine started to cut out. But… I couldn't go on

by Johnny Carrington

shooting at him ... he was a dead duck. He couldn't get away, but... I couldn't bring myself to finish him off. He was in the middle of the Channel by then, and I imagined he'd have a long swim for home if he didn't drown. So, I flew alongside him and wished him good luck... then turned round and headed home. Later I found out he was picked up and actually shot more of ours down so I should have shot him down but... humanity came into it. I just couldn't do it. I had a letter from his wife after the War saying 'if it wasn't for you I wouldn't have my grandchildren'.

Ensemble sing[3]: There's a blue sky driving all the dark clouds away
And a sudden act of kindness changes play
The giving of a life in the throes of wartime strife
To someone that he'll never meet again some sunny day

The whole world turns clown and paints itself red;
Soldiers brave were all too often brought home dead.
Ne'er before was so much owed by so many to so few
The few were near three thousand young and true.

This chink of light lets us feel that all's not wrong,
And gives us all good reason to sing this song:

There's a blue sky driving all the dark clouds away
When a random act of kindness changes play:
The giving of a life in the throes of wartime strife
To someone that we'll never meet again... when... some sunny day.

---

[3] Score on Page 84; PDF of score in downloadable files

Gordon:	But it wasn't just pilots who displayed immense courage during the Battle of Britain. John McKeown was a young apprentice at the Supermarine factory in Woolston. He was fifteen years old and hadn't been there long. He lived in Fawley, a small village a few miles away.
John's Dad:	Southampton has taken a bit of a pasting today love… look at the smoke.
John's Mum:	I hope John is O.K.
Dad:	I'm sure he is.
Gordon:	But later that evening their son had still not returned.
Mum:	No sign of John yet?
Dad:	No love.
Mum:	He should have been back a long time ago.
Dad:	*(Looking out)* I know…
Mum:	I'm worried… just look at Southampton… a massive red glow… John could be in that!
Dad:	There's no use in worrying love. We can't do anything now. If he isn't here in the next hour, I'll go and look for him.
Mum:	You can't go into town.
Dad:	What do you want me to do?
Mum:	I don't know…
Dad:	Come here love.
	Sound of John entering
Mum:	John! *(Rushes over and embraces him)*. Thank God you're back!

by Johnny Carrington

Dad: Are you alright son?... John?... John are you O.K?

John is very quiet

Mum: It's alright my love you're home now. We were so worried. Let me make you some supper.

John: Mum I'm not hungry.

Dad: Cup of tea?

John: No dad.

Dad: Were you caught up in the bombing?

John nods but can't talk

Mum: I'll go and sort your bed out. Your sister has been worrying about you.

Dad: Do you want to talk son?

John shakes his head.

Are you sure?

John looks up

John: Dad... it was dreadful. *(He now begins to cry and let it out).*

Dad: O.K. my lad. Tell me what happened.

He slowly leaves his father and begins to act out the story in another area of the stage.

John: Everything seemed normal to begin with. I was just cleaning the tools with Doug. He isn't as old as me so I was showing him what to do, then we heard the air raid siren go off.

Doug: Come on John, quick.

John: I'm just coming Doug. Quick grab those tools.

Doug:	Come on mate, we don't have time to get them.
John:	We all began to make our way out of the building when I heard the anti-aircraft guns start to go off.
	SFX: Guns
	Blimey those guns are loud!
Doug:	Let's hope they bag a Jerry eh?
John:	Not half. We need to get to the shelters quick. Have you seen Ken?
Doug:	No, let's just get a move on…
	SFX : Sudden explosion, carnage everywhere
John:	*(staggering to his feet)* Doug? Doug? Where are you?
	He sees a body
	Doug??
	(He realises that his friend is dead. Quietly)
	Bloody hell.
Worker:	Quick John, I need a hand over here.
John:	What?
Worker:	Over here… we need to get these people out. They're trapped.
John:	But what about Doug?
Worker:	You can't help him now son…
	Another big explosion
Worker:	Are you alright?... John, are you alright?
John:	… Yes

by Johnny Carrington

Worker: Quick then lad, we need you over here. I'm sorry, but we need to move these bodies.

They start to move bodies and helping injured. John then goes back to the area with his father

John: It was dreadful dad... just dreadful.

Dad: Come on... you're home now. Let's get you up to bed.

Gordon: The following morning John's mother went in to bring him a cup of tea... expecting to find him in bed.

Mum: What are you doing love?

John: Getting ready for work.

Mum: Do you have to go in today?

John: I've got to.

Mum: But after yesterday... they won't...

John: Mum... I've got to be there for 7.30. I don't want to be late

Mum: Are you sure love?

John: They need me more than ever today. We have to be building Spitfires mum. The longer I'm not at work, the longer it will be before we start making them again.

Gordon: And so John McKeown, aged only fifteen, was back at work the day after pulling his dead friends from the rubble that was Supermarine. A day later, the bombers returned.

SFX: Air raid siren etc. over the top of the following dialogue

John: Come on Ken run.

Ken: Not again. Quick, head for the shelters.

John: Which one? There... number thirteen. That'll do.

Ken: No it's unlucky, go to number eleven. Run!

They freeze

Gordon: So they ran... Number eleven took a direct hit. Ken was killed instantly and John was injured and taken to hospital. They thought he'd make a full recovery so he was sent home, but complications set in and six weeks later... he died. None of this story would have been known had his sister not taken John's old scrapbook to the Solent Sky Museum and told Alan Jones about it. It's not a particularly interesting scrapbook, just a few pictures, but the story behind its owner is extraordinary. And today, that is all that is left to remind us of John McKeown, a truly remarkable fifteen year old boy... and now one of the museums most treasured possessions.

Soundscape

Bob: We were scrambled against a bunch of fighters coming in at about 20,000 ft. We went into cloud and I lost everyone. Funny thing cloud... when you are looking down into it you can see an aeroplane climbing out, if you are in the cloud you cannot see a blind thing until you are totally clear. There were twenty 109s above the cloud watching me, rubbing their hands together and boy did they hit me from in front and behind. The bullets went through me, hitting my shoulder, my hand and a cannon shot which hit my parachute. When I was hit in the shoulder I thought I was dead – it was a monumental blow. I had trained myself that if anything came from behind, I would hit the stick and go down because

by Johnny Carrington

that's the one direction you can move quickest... it took me out of the bullet stream. I then had to get out the aircraft, so I pulled my Sutton harness but it wouldn't release – I kept pulling and eventually I just... fell out. Falling through the air without your parachute open is the most lovely feeling – I was in cloud and with my one good arm I pulled the ripcord and it worked. There was a beautiful blue lagoon below me – Poole Harbour and Brownsea Island. There was a quagmire in the middle of the Island and I hit it which gave me a soft landing ... and that was me out of the battle... at least for a few months... that probably saved my life.

Section 8: Success... and tragedy

Old film of the battle can accompany this section. The options are there to use narrators, multimedia or both.

Nar 1: The Battle had raged across Southern England all summer. It was a close-run thing, but against all odds the RAF prevailed and the German invasion plans were cancelled.

Nar 2: Spitfires and Hurricanes flown by British, Commonwealth, Polish, South African, American and French pilots shot down nearly 2000 German aircraft but lost 537 pilots in the process.

Nar 3: For the first time the mighty German war machine had been stopped, and although it would be another five years before the allies would win the war, Germany's myth of invincibility had been broken.

Nar 4: Broken by a small band of men who flew their fighters bravely and supported by people on the ground who risked their own lives to make sure that these pilots had the machines they needed to fight.

Multimedia 11

Nar 2: Today, all these years after The Battle of Britain, the Spitfire is still being flown and tenderly restored by organisations such as 'The Fighter Collection' based at the old Battle of Britain station in Duxford.

Nar 3: Here their purpose is far less dangerous. They swoop and soar to the delights of many air show visitors, from all nations, and they still have the power to inspire…

Nar 4: In some ways history has been kinder to the Spitfire than the Hurricane… after all in the Battle the Hurricane shot down more aircraft. But even before the war the Spitfire held the public's imagination like no other aircraft. It was faster, more elegant than anything else the British had.

Narrator 1: If the Hurricane was the workhorse… dependable and sturdy, the Spitfire was the racehorse… a fast glamorous thoroughbred. It must have been special because it was the only allied fighter to remain in production through the whole of the war.

Gordon: Bob Doe sums it up nicely when saying:

Bob: Spitfire – to fly it was a dream it had no vices but it had the ability to fight anything on earth – it was the most wonderful aeroplane. If you are at an air show and a Spitfire flies over the whole place stops to look

by Johnny Carrington

	don't they... everyone stops to listen to a Spitfire... so distinctive... a wonderful sound.
Gordon:	My father would never have thought his design would be still flying all these years after its maiden flight... 1936 had been a very busy year. His trusted employee Joe Smith would continue to develop the Spitfire, and my father would busy himself with his next design, and his own flying, and occasional sailing... it was what kept him going... for he could feel his illness was returning... with a vengeance.
Gordon:	Whilst I was at boarding school I received the following letter almost a year to the day after the Spitfire first flew: it typified the fortitude with which my father dealt with his illness.
RJ:	*(Reading)* My Dear Gordon, Very many thanks for your letter. We were very pleased to hear from you to know you had arrived safely. To answer your question Gordon, I think it would be a good idea to stay another term, especially as you would like to, and as our arrangements have been messed around by my inconsiderate behaviour.

The narration of the letter switches to younger Gordon

Young Gordon:	About your visit next weekend Gordon, you know how delighted I would be to see you, and yet I suppose there is a limit to these things. So far we have been perfectly justified in what we have done, as at one time I was very, very ill. I tell you what, we will ring you up on Friday evening and then decide whether you come or not.

Narration back to RJ

RJ:	It is snowing heavily this morning and the lawn is covered already-you will be able to make a snowman if you come down. Mum joins me in love to you, Gordon.
RJ / Gordon:	Yours affectionately, Dad.
Gordon:	Then one day, my mother found something that gave them renewed hope.
	Flo is in the sitting room flicking through a magazine. RJ is in his normal chair asleep.
Flo:	Reg? *(RJ begins to stir)*. Reg... Reg there is an article in here that talks about a revolutionary new treatment for cancer. *(RJ, who had been half-following what Flo said, suddenly becomes more alert)*. It says here the new treatment is drastically increasing survival rates. It's in Vienna. Oh Reg, we need to try this, even if it means travelling to Austria. What do you think?
RJ:	*(Smiling)*. Why not?
Flo:	Excellent. I'll start enquiries in the morning. I'm going to phone Gordon's school, I want to tell him.
RJ:	Steady Flo, let's not get ahead of ourselves yet love... all we've seen is a magazine article.
Flo:	I know, but this is a chance. It's a chance Reg... we've got to be positive.
	Transition music
Gordon:	A plane journey to Vienna was followed by a taxi ride to the hospital. A Doctor Smitt was waiting for them
Dr Smitt.	Mr and Mrs Mitchell, good afternoon.
RJ/Flo	Hello/how are you?

by Johnny Carrington

Dr Smitt:	I think it would be best if we started with a tour of the facilities. Then, as we talked about, start the tests. Quite a few I'm afraid. This will determine if we can proceed. If we decide to go ahead, then we can operate on Thursday.
Flo:	So quickly.
Dr Smitt:	Speed is of the absolute essence Mrs Mitchell. The sooner the better.
Dr Smitt:	My orderly here will show you around, and we will begin tests this afternoon. I hope you don't mind but I have supplied a wheelchair.

Suitable period (pre-1937) music could be used here

It is evening and the Mitchells are getting ready in their hotel room. Both are feeling positive and cheerful. Both are positioned mid-stage standing facing forward

RJ:	You know I feel like a pin cushion.
Flo:	*(Laughing)*. Oh stop fussing Reg, you knew it would happen.
RJ	I'm surprised there's any blood left in me. What time is our reservation?
Flo:	Eight thirty.
RJ:	*(Looking at his watch)*. We need to get a move on.

Restaurant

Flo:	This is nice.
RJ:	What the meal?
Flo:	No... well I mean yes it is... but it's nice us being out. We haven't done it for ages. I remember the first time

	you took me out... on the back of your bike! Do you remember me holding on for dear life, it was a complete death trap
	They laugh
RJ:	Ha! I had forgotten all about that. Crikey we had some near misses.
Flo:	Now you tell me.
RJ:	You must have seen?
Flo:	You're joking, I always had my eyes closed.
RJ:	I don't know how we didn't kill ourselves on the back of that thing.
Flo:	You were driving!
RJ:	You were egging me on.
Flo:	I was not!
RJ:	You were.
Flo:	I was a respectable Head Mistress... I would have done no such thing... well maybe a bit. *(They both laugh)*. What did we see on our first date then?
RJ:	Umm... can't remember.
Flo:	Oh Reg!
RJ:	Of course I can... Chaplin.
Flo:	*(Pause)* Those days were fun.
RJ:	They certainly were. If this operation is successful, I was thinking we could start playing tennis again.
Flo:	One thing at a time maybe... but yes, that would be lovely... .

The next morning there is the sound of them getting out of a taxi at the hospital. They make their way to the doctor's office and wait outside.

Secretary: Dr Smitt will see you now

RJ / Flo: Thank you.

Dr Smitt: Good morning, have a seat.

Flo: Good morning doctor. We have been chatting about what we might do when we are back in England and we...

Dr Smitt: Mrs Mitchell.

Flo: Yes?

Dr Smitt: I'm sorry to say... the cancer has spread too much I'm afraid. The treatment just wouldn't work.

Everyone is quiet for a moment.

Flo: But Reg has the strength to fight this. He has been for the last three years.

Dr Smitt: I'm sorry but there...

Flo: We've got more money

Dr Smitt: It's not a question of money

Flo: *(With desperation creeping into her voice)* How much? Whatever it costs, we can pay for it

Dr Smitt: Mrs Mitchell, I'm sorry, but no amount of money will make any difference. The cancer has spread too far.

Flo: But you must be able to do something?

Dr Smitt: I'm so sorry.

	She just turns to RJ and starts to sob. The doctor, without saying anything, quietly gets up and leaves.
RJ:	Come on Flo… let's go home.
Gordon:	I think everyone was just resigned to the inevitable after that. There was never any more talk of cures… My mother wanted me to know the truth and severity of the illness.
Flo:	Gordon, I need to talk to you my love… It's about Dad. He really isn't well.
Young Gordon:	I know… I thought it was bad this time. I could tell.
Flo:	It really is very bad. The treatment we have tried… it just hasn't worked.
YG:	But they can't have done everything?
Flo:	The Doctors have tried so very hard my love. But they just can't cure it.
YG:	Is he going to… to…
Flo:	*(Nodding).* I'm afraid he is my love. I am so sorry. *(Holding him).*
YG:	*(Pulling away)* Surely there must be something though?
Flo:	There isn't Gordon.
Gordon:	Dad would spend most of his time in the garden doing what he could or feeding the fish in the pond. It was sad to see him so… so quiet and virtually immobile after what had been such a dynamic life. I was allowed home from school when it seemed that… well… that the end was near.

by Johnny Carrington

SFX: Spitfire overflies

Flo: Here's your tea I bet you knew when we bought this house Reginald Mitchell, we would end up on the flight path.

RJ: As if I would?

SFX: Front door bell

I wonder if that's Gordon?

Flo: I'll just put this on the table and see.

She places the cup on the table next to RJ and exits back into the house. Muffled voices can be heard from inside the house.

YG: *(Entering)* Hello dad. How are you feeling?

RJ: Oh you know... It's good to see you son. How was your journey?

YG: Oh pretty good. *(RJ groans as he stretches for his cup of tea).* I'll get that for you dad. *(Gordon reaches for the cup and hands it to RJ).*

RJ: Thanks. It's pathetic isn't it... I can't even get my own cup of tea...

YG *(Pause)* The garden's looking good.

RJ: Could do with a bit more rain.

YG: I hear the Spitfires are about to go into squadron service.

RJ: Not far off. A couple of snags with icing in the guns... but nothing that won't take too long to sort out.

YG: They're saying that it will be more than a match for Willy Messerschmitt's one-oh-nine.

RJ:	I'm not sure about that... but I think it will hold its own.
YG:	Well that's what they're saying.
	Suddenly there is a smash from the kitchen.
	Are you alright mum?
Flo:	*(From kitchen)* It's just a plate!
RJ:	Go and check on her son, it's been happening rather a lot recently.
	Gordon goes to the kitchen.
Flo:	It's stupid. I keep dropping them.
	Gordon goes over to help.
YG:	Let me do it for you.
Flo:	I'm sorry Gordon.
	He picks them up and as he stands he sees Flo facing away from him with her head lowered. He can see that her shoulders are shaking slightly.
YG:	Mum?... Mum?
	Flo says nothing. Gordon goes over to her and can then see that she is sobbing.
	Mum?
	He turns her around and holds her. They say nothing for a while but Gordon just holds her.
	There's nothing to be sorry about.
Flo:	But this is the last thing you need. I really didn't want to be like this.

by Johnny Carrington

YG:	Mum, it doesn't matter... It must be difficult for you... looking after dad by yourself. *(She nods and begins to sob again).* I'll make the tea. *(Flo smiles and wipes her eyes).*
Flo:	Sorry.
YG:	Mum, stop saying that. What you have done for dad is nothing short of remarkable. Go on... go and sit with him. I'll bring it out.
Flo:	*(Flo lets out a sigh).* Thanks Gordon. *(She leaves the kitchen and goes outside. She sits down next to RJ).* Hello Love.
RJ:	Hello... Thanks Flo.
Flo:	What for?
RJ:	*(Long pause. He looks at Flo, he reaches out and holds her hand)* Everything... absolutely everything.
	SFX: Another Spitfire flies overhead
	Multimedia 12 (RJ Mitchell) could be inserted at any moment during the following monologue
Gordon Mitchell:	It was so difficult to come to terms with. How does a young lad do that?
	RJ, Flo and YG on stage during this monologue.
	The last three months of my Father's life were very difficult. He was racked with pain which not even morphine could dull. My mother would sit with him in the garden, him wrapped up in a big overcoat, and talk of trivial things like the dog or his pond, as if nothing was wrong. We lived close to the airfield and I imagine him hearing the sound of the Spitfire passing overhead, and wanting to be at work, but he

was too weak... I cannot emphasise too strongly what I know from first-hand experience, my father displayed the greatest courage and fortitude in the face of severe physical adversity.

Then, on June 11th 1937, he passed away. Although we had both been expecting it, it seemed impossible to accept. I just felt numb... and very sad. It was devastating to have my father taken away just when I began to need his help and advice most, but even at the tender age of 16½, I understood that at least he was no longer in pain.

The funeral service at Highfield Church was beautiful and very moving... a lone Spitfire flew overhead.

SFX: *Merlin engine sound*

He was never to see it help to save our nation, never saw its finest hour... I often think that was somehow unfair. When my mother died in 1947, her remains were interred with his.

He was a good, caring father. Like us all he had his faults, but he was someone who taught me to stand on my own two feet.

I often wonder what he thought flying the Spitfire was like. I think back to that poem by Pilot Officer Magee.

The poem is repeated with one character speaking the first two lines and more characters joining in as it progresses until the whole cast is speaking.

Oh I have slipped the surly bonds of earth

And danced the skies on laughter-silvered wings:

Sunward I've climbed, and joined the tumbling mirth

Of sun-split clouds - and done a hundred things

You have not dreamed of - wheeled and soared and swung

High in the sunlit silence. Hov'ring there

I've chased the shouting wind along, and flung

My eager craft through footless halls of air.

Up, up the long delirious, burning blue,

I've topped the windswept heights with easy grace

Where never lark, or even eagle flew -

And, while with silent lifting mind I've trod`

The high untrespassed sanctity of space,

Put out my hand and touched the face of God.

SFX: *The sound of a Spitfire approaching, flying directly overhead and then receding into the distance*

As this happens, to fit the sound, the cast look at the approaching Spitfire which is coming from the direction of the audience, look straight up as it flies overhead and turn to look at it as it flies away, waving at it.

The lights dim with the cast facing away from the audience, waving, so that they are silhouetted until the lights

FADE TO BLACK

Multimedia Files

There are seventeen specially created multimedia files which can be used in the performance of this script.

Most are for projection and accompany the onstage action. One is a sound file of the song *Some Sunny Day*. The score as a PDF is also included. One is a recording of Churchill's "We shall fight them…" speech.

Some of the files contain contemporaneous footage (the publisher is grateful to the RAF Museum Film and Sound Archive) and have sound, are complete in themselves and form part of the action. Others have no sound and are projected during onstage dialogue (and can be faded when the dialogue is finished as they are deliberately over-long).

There is also a blank "template" if you need to add any additional projections for your own production.

To access the files type this link into your browser:

https://bit.ly/mitchellswings

Multimedia 1 in the script is labelled MM1 in the files, and so on.

All of the files are copyright and are made available to you solely for use when working on and presenting this play.

If you have any problem accessing the files or have any queries relating to this play then please contact the author via

johnnycarringtonpublishing@gmail.com

Also by Johnny Carrington

Bang Out Of Order
(with Danny Sturrock)

4 friends, 1 secret, 1 chance, 1 life.
A play which tackles anti-social behaviour head on. This roller-coaster ride will educate, amuse and challenge.
Suitable for: Key Stage 3 to adult
Duration: 50 minutes approximately
Cast: 4 female, 4 male, or 2 female and 2 male with doubling.

Jamie in the Land of Dinnersphere
(with Mark Wheeller)

Jamie is an ideal project for vocational courses; KS4 students perform to KS2.
Jamie Jamjar loves healthy food. He has seen how a poor diet can mess you up... Jamie is shocked when his school replace the friendly dinner ladies with Robot Dudes (fast food servants). Then he discovers his own father invented them and embarks on a mission to Dinnersphere to improve the situation.
Suitable for: Performance by Key Stage 4 to Key Stage 2 audience
Duration: 35 minutes (50 minutes with workshop)
Cast: 1 female, 3 male, 5 or 6 female/male, or 2 female and 2 male with doubling.